# Mitigation of Metal Toxicity in Plants

Riaz Ahmad
https://orcid.org/0000-0002-4383-8335
*The University of Agriculture, Dera Ismail Khan, Pakistan*

IGI Global
Scientific Publishing
Publishing Tomorrow's Research Today

| | |
|---|---|
| Vice President of Editorial | Melissa Wagner |
| Director of Acquisitions | Mikaela Felty |
| Director of Book Development | Jocelynn Hessler |
| Production Manager | Mike Brehm |
| Cover Design | Jose Rosado |

Published in the United States of America by
IGI Global Scientific Publishing
701 East Chocolate Avenue
Hershey, PA, 17033, USA
Tel: 717-533-8845 | Fax: 717-533-7115
Website: https://www.igi-global.com E-mail: cust@igi-global.com



Library of Congress Cataloging-in-Publication Data

Names: Ahmad, Riaz, 1993- editor http://id.loc.gov/rwo/agents/no2025022058
Title: Mitigation of metal toxicity in plants / edited by Riaz Ahmad.
Description: Hershey, PA : IGI Global Scientific Publishing, 2026. | Includes bibliographical references and index. | Summary: "This book strives to reduce the production losses in crops subjected to metal toxicity, improve plant defense system against metal toxicity, cover morphological, physiological, and biochemical adjustments under metal toxicity, and improve plant tolerance against metals toxicity by modulation of anatomical changes and metabolites generation"-- Provided by publisher.
Identifiers: LCCN 2025007446 (print) | LCCN 2025007447 (ebook) | ISBN 9798369398999 hardcover | ISBN 9798369399002 paperback | ISBN 9798369399019 ebook
Subjects: LCSH: Plants--Effect of metal on | Plant ecophysiology http://id.loc.gov/authorities/subjects/sh88005635 | Plants, Protection of http://id.loc.gov/authorities/subjects/sh85103025
Classification: LCC QK753.M47 M58 2025 (print) | LCC QK753.M47 (ebook) | DDC 571.2--dc23/eng/20250502
LC record available at https://lccn.loc.gov/2025007446
LC ebook record available at https://lccn.loc.gov/2025007447

British Cataloguing in Publication Data
A Cataloguing in Publication record for this book is available from the British Library.

# Table of Contents

# Detailed Table of Contents



Crop yields are declining considerably around the globe due to heavy metal contamination. Higher concentration of metals not only hinders plant development, lowers harvest and its quality, but consumption also poses a health risk for humans. Soil pollution is primarily caused by urban expansion, industrialization, polluted sewage sludge, and overuse of chemical fertilizers. Moreover, the increase in toxic metals is due to the use of untreated wastewater for crop irrigation. The accumulation of these toxic metals can disrupt the physiological functioning of the plant and adversely affect its overall performance. The exposure to metal toxicity can disrupt nutrient uptake and reduce photosynthetic activity. This also leads to oxidative harm and metabolic disturbances. The consumption of contaminated food is injurious to health. Therefore, it is crucial to address this challenge, and some of the management strategies that are being employed to combat metal toxicity are bio-stimulants, phytohormones, soil microbes, biochar, and selection of resistant varieties.



Climate change has disrupted farming practices due to abiotic stress conditions. Metal toxicity is a major threat to agriculture. The drastic growth reduction, poor yield, and inferior quality were recorded from the metal toxicity. Metal toxicity contributes to the impairment of food safety and exhibits adverse effects on crop production. Plant growth regulators (PGRs) are well-known because of their excellent contribution to the mitigation of the adverse effects of metal toxicity occurring in plants. The antioxidant defense system of plants is boosted with the exogenous supplementation of PGRs. Different aspects, such as uptake, translocation, and detoxification of metals, can be regulated with PGRs supplementation. The harmful effects of metals on plants were significantly mitigated with exogenous PGRs supplementation. Phytohormone use is a long-term approach to reducing the adverse effects of metal stress conditions. PGRS are an excellent opportunity for the mitigation of adverse effects of metal toxicity in plants by improving growth, yield, and quality.



Climate change is considered as one of the major constrain reducing the productivity of crops. The soil contamination is increasing gradually due to the several factors such as brackish water, urbanization, and industrial effluents. Phytoremediation is considered as effective remedy for the removal toxic substances form the soil focusing on sustainable growth and higher yield. However, current chapter aimed to evaluate the impact of biostimulants in mitigation of metal toxicity in plants. Biostimulants are widely used in agriculture to increase crop yield and resilience against various stresses. Moreover, recent research has shown that biostimulants can lessen the harmful effects of contaminants on plants, boosting some species' phytoremediation effectiveness. Biostimulants composed of humus, protein and amino acid hydrolysate, inorganic salts, microbes, seaweed, plant extracts, and fungi are used on crops against metal toxicity. Conclusively, biostimulants are effective way to cope adverse effects of metal toxicity in plants.



Nitric oxide (NO) is a redox related signaling molecules that is thought to be important regulators involved in plant growth and various developmental processes as well as the way plants react to abiotic stress conditions. Metal toxicity causes reactive oxygen species (ROS) to be produced, which brings plants under oxidative harm. Plants use various ROS-scavenging systems, including signaling molecules, non-enzymatic antioxidants and antioxidative enzymes, to reduce the negative moments of metal toxicity. NO has high potential to mitigate the adverse effects of metal toxicity by regulation of ROS overproduction. The way in which nitric oxide triggers the proper set of reactions to metal toxicity requires a comprehension of its functions. In this chapter concentrate on the most recent information regarding nitric oxide's mitigating contribution to metal toxicity,encompassing everything compared to metal toxicity,build-up to the antioxidant response to associated gene and protein regulation.



Heavy metal deposition in soils can negatively impact crop growth and development owing to phytotoxicity, food safety and marketability, and the environmental health of soil organisms. It has become an issue in agricultural production. Plants' metabolic processes have an impact on the biological and geological dispersion of heavy metals by contaminating the air, water, and soil. Plants are an essential part of the ecosystem. Metal buildup, reduced development, altered metabolism, and decreased biomass production are all observed in plants living in metal-polluted environments. Metals have an impact on many physiological and biochemical processes in plants. Plant responses to metal toxicity and plant growth and development are regulated by jasmonates (JAs), which are widely distributed lipid-derived chemicals. Jasmonic acid belongs to the oxylipin family, as are its several metabolites. Hence, it has been illustrated that jasmonic acid plays a significant role in improving crops' tolerance to physiological and biochemical mechanisms that lessens the adverse effects of metal toxicity.



Metals at optimum level are necessary for plants and human heaths. However, their excess is dangerous for living organisms. Heavy metals can alter the permeability of mitochondrial membranes and limit the activity of reactive oxygen species (ROS) clearance enzymes in the cellular antioxidant system. However, they cause plants to produce excessive amounts of ROS. The buildup of environmental contaminants and oxidative stress in exposed species is caused by the interaction of ROS with cellular membranes, heavy metal-induced interactions with various macromolecules either directly or indirectly, and signaling pathways. Numerous pathogenic processes, including energy depletion, excessive ROS production, mitochondrial respiratory chain damage, decoupling of oxidative phosphorylation, and mitochondrial death, are obstructed by the heavy metal–ROS–cell signaling axis. Salicylic acid had excellent potential mitigates adverse effects of metal toxicity by regulation of plant growth, photosynthetic pigments, and antioxidant defense system in plants.

The stunted growth, poor nutrient uptake, oxidative damage, and reduced physiological functioning are symptoms of metal toxicity in plants. Biochar, a stable carbon-rich substance produced by pyrolyzing organic materials in an oxygen-limited environment, is a potential method for enhancing soil quality under metal stress. Its large surface area, porous structure, and capacity to interact with soil constituents, biochar can improve soil fertility, increase water retention, and lower the bioavailability of harmful metals. Biochar promotes plant growth and developmental processes and increases tolerance to metal toxicity. It explores the fundamental processes by which biochar interacts with soil and plant systems, enhances plant defenses, controls photosynthesis, increases plant growth, and reduces the harmful compounds produced under metal stress. This chapter attempts to validate the potential of biochar as an environmentally sustainable method to reduce the negative effects of metal contamination in agriculture.



Abiotic stress conditions occur from climate obstacles to crop productivity. Among these, metal toxicity in progressively increasing and polluting the soils. Current aimed to explore the adverse effects of metals on plants. It explores the sources of these metals, examining both natural occurrences and human-induced activities, and investigates the mechanisms through which plants absorb them. Metal pollution negatively affects plant and microbiome well-being, producing reactive oxygen species (ROS) that harm essential macromolecules which further contributes in various developmental processes. The need for significant development in conventional stress-resistant plant cultivars has prompted the investigation of novel strategies such as phytohormones, biostimulants, organic amendments, and nanotechnology. Nanoparticles have significant impact on alteration of morphological and biochemical properties of normal and stressed plants.Long-term ecological studies and customized

# Preface

## INTRODUCTION

The accelerating pace of industrialization and urban expansion has left an undeniable imprint on our natural ecosystems, most notably through the accumulation of heavy metals in soil and water. As these elements infiltrate agricultural systems, their entry into the food chain poses a serious challenge not only to plant productivity but also to environmental sustainability and human health. This volume, *Mitigation of Metal Toxicity in Plants*, has been carefully compiled to address these pressing concerns with both scientific rigor and practical relevance.

Plants exposed to elevated concentrations of heavy metals experience profound disruptions in their physiological and biochemical processes. These disturbances often manifest as reduced seed germination, stunted seedling growth, diminished biomass, and compromised metabolic functions. Recognizing these challenges, researchers across disciplines have developed and refined a range of management strategies aimed at alleviating the adverse effects of metal toxicity.

This edited reference work brings together current knowledge on such mitigation approaches, including the application of phytohormones, natural plant extracts, beneficial microbes, water purification techniques, and organic amendments. Collectively, these strategies contribute to strengthening plant defense systems, enhancing tolerance mechanisms, and enabling crops to maintain functional stability under stress conditions.

The objectives of this volume are multifold: to explore ways to reduce production losses in crops affected by metal toxicity; to deepen understanding of plant defense mechanisms; to examine morphological, physiological, and biochemical adaptations; to highlight anatomical and metabolic modulations that improve tolerance; and to identify and characterize genetic resources capable of withstanding metal stress.

This book is intended to serve a broad audience, including professionals in horticultural trades and industries, academic scholars, researchers, students, farmers, plant breeders, and personnel at research stations. It will also be of value to stakeholders in crop export sectors, food industries, and chemical-related trades, all of whom are increasingly impacted by issues of environmental contamination and crop resilience.

As editor, I extend my sincere appreciation to the contributors whose expertise and dedication have made this compilation possible. I hope that this work will not only advance scientific understanding but also support the development of sustainable agricultural practices in environments challenged by metal toxicity.

## ORGANIZATION OF THE BOOK

Chapter 1 provides a comprehensive examination of the adverse effects of metal toxicity on crops by alteration of physiological and biochemical indices. Crop yields are declining considerably due to metal contamination. Higher concentration of metals not only hinders plant development, lowers harvest and its quality, but consumption also poses a health risk for humans. Soil pollution is primarily caused by urban expansion, industrialization, polluted sewage sludge, and overuse of chemical fertilizers. The increase in toxic metals is due to the use of untreated wastewater for crop irrigation. The accumulation of these toxic metals can disrupt the physiological functioning of the plant and adversely affect its overall performance. Some management strategies, i.e., bio-stimulants, phytohormones, soil microbes, biochar, and the use of tolerant cultivars, can combat the metal toxicity.

Chapter 2 focuses on the pivotal role of plant growth regulators in enhancing plant tolerance to metal toxicity under changing climatic conditions. It elaborates on how exogenous application of phytohormones strengthens antioxidant defense systems, regulates metal uptake and translocation, and improves overall plant growth, yield, and quality. The chapter positions plant growth regulators as a sustainable and long-term solution for mitigating the adverse effects of heavy metal stress in agricultural systems.

Chapter 3 explores the application of biostimulants as an innovative and sustainable strategy to alleviate metal-induced toxicity in plants. It discusses the composition of biostimulants, including humic substances, amino acids, microbes, and plant-based extracts, and their role in enhancing plant resilience and phytoremediation efficiency. The chapter underscores the growing importance of biostimulants in improving crop productivity and stress tolerance under contaminated soil conditions.

Chapter 4 examines the role of nitric oxide as a critical signaling molecule in mitigating the harmful effects of metal toxicity. Emphasis is placed on its ability to regulate reactive oxygen species and activate antioxidant defense mechanisms. The chapter synthesizes recent advances in understanding nitric oxide-mediated signaling pathways, including gene and protein regulation, offering insights into its central role in plant adaptation to metal stress.

Chapter 5 delves into the role of jasmonates in regulating plant responses to heavy metal toxicity. It highlights how these lipid-derived signaling molecules influence morphological, physiological, and biochemical processes, thereby enhancing plant tolerance. The chapter provides a detailed account of how jasmonic acid and its derivatives modulate stress responses and contribute to maintaining growth and metabolic balance under toxic conditions.

Chapter 6 presents an in-depth analysis of salicylic acid as a key regulator in plant defense against metal toxicity. It explains how salicylic acid modulates physiological and biochemical pathways, including antioxidant activity, photosynthesis, and cellular signaling, to counteract oxidative stress. The chapter emphasizes its role in maintaining cellular integrity and improving plant resilience under heavy metal exposure.

Chapter 7 evaluates the potential of biochar as a sustainable soil amendment for mitigating metal toxicity. It discusses the physicochemical properties of biochar, such as its porous structure and high adsorption capacity, which reduce metal bioavailability and improve soil fertility. The chapter also highlights biochar's role in enhancing plant growth, regulating physiological processes, and strengthening defense mechanisms under stress conditions.

Chapter 8 investigates the emerging role of nanotechnology in improving plant tolerance to metal toxicity. It examines the sources and impacts of metal contamination, as well as the interaction between nanoparticles and plant systems. The chapter emphasizes how nanotechnology can modify morphological and biochemical traits, offering innovative solutions for stress management, while also addressing the need for long-term ecological assessments.

## CONCLUSION

In bringing together the diverse contributions presented in this volume, it becomes evident that the challenge of metal toxicity in plants is both complex and multifaceted, requiring equally dynamic and interdisciplinary solutions. The chapters collectively underscore that heavy metal stress is not merely a physiological constraint but a broader environmental and agricultural issue that demands integration of molecular insights, ecological understanding, and practical interventions.

This work highlights significant progress in unraveling the mechanisms by which plants perceive, respond to, and tolerate toxic metal exposure. From hormonal regulation and signaling molecules to soil amendments, biostimulants, and advanced nanotechnological approaches, the breadth of strategies discussed reflects the evolving nature of research in this field. Equally important is the emphasis on sustainable and environmentally responsible practices, which are essential for long-term agricultural resilience.

The insights compiled herein reinforce the importance of bridging fundamental research with applied science. The identification of tolerant genetic resources, the refinement of mitigation techniques, and the advancement of innovative tools such as biochar and nanoparticles collectively offer promising pathways to safeguard crop productivity and food security in contaminated environments.

As editor, I am confident that this volume will serve not only as a reference for current knowledge but also as a catalyst for future research and collaboration. We hope that the ideas and findings presented will inspire continued efforts toward developing effective, sustainable solutions to mitigate metal toxicity in plants, ultimately contributing to a more resilient and secure agricultural future.

# Chapter 1
# Detrimental Effects of Metal Toxicity on Crop Growth, Productivity, and Quality Attributes

**Ahmad Hassan**
*Huazhong Agricultural University, China*

**Ans Mujahid**
*Bahauddin Zakariya University, Multan, Pakistan*

**Muhammad Adnan Sadique**
http://orcid.org/0000-0002-4383-8335
*Bahauddin Zakariya University, Multan, Pakistan*

## ABSTRACT

*Crop yields are declining considerably around the globe due to heavy metal contamination. Higher concentration of metals not only hinders plant development, lowers harvest and its quality, but consumption also poses a health risk for humans. Soil pollution is primarily caused by urban expansion, industrialization, polluted sewage sludge, and overuse of chemical fertilizers. Moreover, the increase in toxic metals is due to the use of untreated wastewater for crop irrigation. The accumulation of these toxic metals can disrupt the physiological functioning of the plant and*

DOI: 10.4018/979-8-3693-9899-9.ch001

*adversely affect its overall performance. The exposure to metal toxicity can disrupt nutrient uptake and reduce photosynthetic activity. This also leads to oxidative harm and metabolic disturbances. The consumption of contaminated food is injurious to health. Therefore, it is crucial to address this challenge, and some of the management strategies that are being employed to combat metal toxicity are bio-stimulants, phytohormones, soil microbes, biochar, and selection of resistant varieties.*

## INTRODUCTION

Urbanization and industrial effluents are major threats to fertile lands with the accumulation of metals in excess. The dependency on the usage of contaminated wastewater for irrigation is mainly attributed to the scarcity of freshwater (Mishra, 2023). This practice has increased globally over the past two decades, in both developed and developing states. The wastewater contains both essential plant nutrients and organic contaminants (Bijekar et al., 2022). According to Qadir et al. (2010), the majority of the farmers, especially the poor ones, use wastewater for irrigation to enhance crop productivity as it is nutrient-rich and an alternative to costly fertilizers. Under certain conditions, peak crop yields can be obtained by using wastewater (Jingjing and Xiaohua, 2022).

Pakistan, being one of the major agricultural states, relies on the Indus basin canal system for its irrigation purposes (Basharat, 2019). Despite that, an unpredictable surge in population and climate change has rendered the canal water insufficient to meet the agricultural needs of the country. Consequently, the use of untreated wastewater is widespread, and no serious effort has been made by the regulatory authorities to cope with this harmful and dangerous practice (Dickin et al., 2016). This trend has contributed to increasing levels of food contamination, biodiversity loss, and public health crises. Untreated wastewater is a leading cause of pediatric infections and is directly linked to outbreaks of cholera, typhoid, hepatitis, dysentery, vomiting, diarrhea, and chronic skin allergies. Moreover, lethal body illnesses, such as high blood acidity, cancer development, and stunted growth, are directly associated with metal poisoning (Engwa et al., 2019).

Farmers who use wastewater also face significantly higher health risks and are more vulnerable to toxicity (Woldetsadik et al., 2018). Plants have served as the key source of food for mankind since the earliest times, and consuming food from contaminated plants has led to heavy metal toxicity in people. Toxic metals like cadmium and lead can accumulate in the human body over time, resulting in metabolic irregularities, kidney dysfunction, bone degradation, neurological disorders, and damage to the central nervous system (Rroji et al., 2022). Although there are risks, some studies have reported positive effects of wastewater irrigation on plant growth. For instance, the research conducted by Hassena et al. (2018) observed improvements in olive tree height, stem diameter, shoot number, and leaf proline content. Similarly, higher fruit yield and total soluble solids have been observed in citrus compared to tubewell irrigation (Paudel et al., 2018). However, excessive accumulation of heavy metals beyond permissible limits was documented in crops such as guava (Yap et al., 2021). Likewise, Grattan et al. (2015) reported elevated levels of boron and magnesium in citrus leaves irrigated with wastewater.

Globally, concerns regarding the metal contamination in vegetables are increasing, with numerous reports highlighting toxicity in various crops (Khanna, 2011). Crop yield is drastically reduced as a result of metal toxicity, irrespective of other stressors (Kalaivanan et al., 2016). The presence of heavy metals in food poses a major risk to human health. This concern and the adverse effect of heavy metal contamination on plants' development and yield are also highlighted by Pretty (2008). This toxicity has been linked to suppressing plant growth in numerous plants, which include cauliflower (Sinha et al., 2015), Chinese cabbage (Mi et al., 2019), onion (Bystricka et al., 2016), broccoli (Mourato et al., 2015; Yeşilyurt, 2023), and pepper (El-Saadony et al., 2024). Similar detrimental effects were reported in radish (Shahzad et al., 2024), *Brassica napus* (Xie et al., 2022), spinach (Vojodi Mehrabani, 2019), and tomato (Piotto et al., 2018). Lettuce exposed to cadmium toxicity showed heightened levels of Malondialdehyde (MDA).

*Figure 1. The adverse effects of metal toxicity on plant growth, yield, and quality*

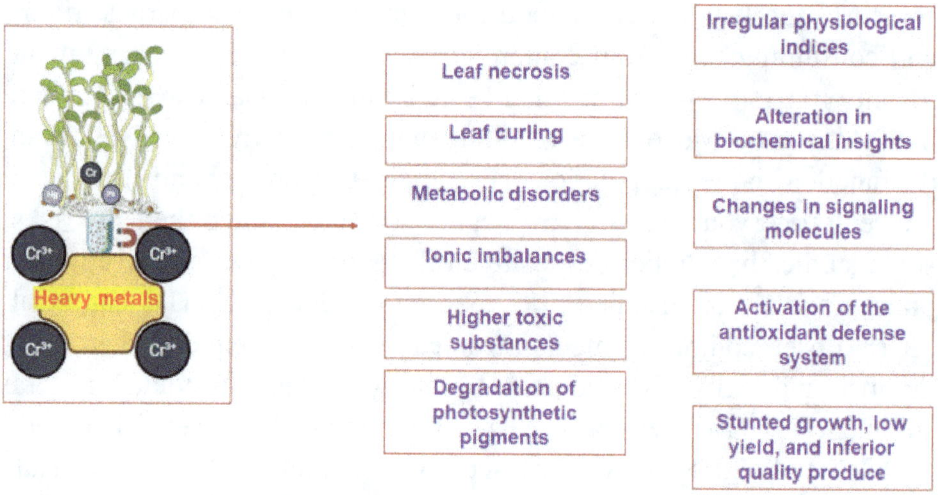


Metal toxicity offers stunted growth and poor productivity of plants (Figure 1). Polluted soils with excess metals are progressively increasing in the future due to urbanization and industrial effluents. Excess of metals translocated from roots to other plant parts (Xie et al., 2022). The consumption of such contaminated products can be toxic to health. Moreover, polluted soils provide irregular photosynthetic pigments, instability in metabolic activities, induced-osmotic stress, excess of toxic substances, resulting in stunted growth and low productivity of crops (Mourato et al., 2015). The current chapter explored the adverse effects of metal toxicity in plants, with an exploration of physiological and biochemical insights involved in the mitigation of metal toxicity.

## MORPHO-PHYSIOLOGICAL AND BIOCHEMICAL CHANGES IN PLANTS UNDER METAL TOXICITY

Metal toxicity has emerged as one of the most hazardous pollutants to mankind and the global ecosystem, with contemporary research pointing out that they are more damaging than pesticides and other common inorganic pollutants such as carbon dioxide and sulfur dioxide (Hassan,

2019). According to Hotamisligil and Davis (2016), plants interact with external stressors, especially heavy metal contamination, through a complex multi-stage mechanism that includes perception of external stress signals, their transmission and integration at the genomic level, and the activation of adaptive responses that modify cellular physiological and biochemical conditions to mitigate the harmful effects of the external stress. However, evaluating these signaling and response mechanisms at the whole-plant level remains challenging. To better understand these signals and alterations, it is useful to examine early responses such as excessive production of reactive oxygen species (ROS), transcriptional and proteomic changes, and metabolite accumulation under stress conditions. For instance, a study conducted by Rizvi and Khan (2017) observed that Cd-stressed plants depicted a considerable reduction in root biomass (up to 73%), primarily due to increased ROS generation (up to 116%) as opposed to control plants. As roots are the first plant organs to come in contact with heavy metal stress, extensive research has focused on their responses for evaluating metal toxicity (Guarino et al., 2018).

*Table 1. Different heavy metals, their resources, and occurrence forms in the soil*

| Metal name | Common source | Available form in soil |
|---|---|---|
| Cadmium (Cd) | Industrial effluents | $Cd^{2+}$ |
| Lead (Pb) | Paints and mining | $Pd^{2+}$ |
| Chromium (Cr) | Tannery waste and industrial effluents | $Cr^{3+}$, $Cr^{6+}$ |
| Arsenic (As) | Groundwater and pesticides | $As^{3+}$, $As^{5+}$ |
| Mercury (Hg) | Industrial discharge | $Hg^{2+}$ |
| Copper (Cu) | Fungicides and fertilizers | $Cu^{2+}$ |
| Zinc (Zn) | Fertilizers, natural | $Zn^{2+}$ |

Soils that are typically polluted with heavy metals inhibit root growth and elongation of plants, which clearly shows that heavy metals restrict overall plant development (Table 1). Numerous research studies have also demonstrated that heavy metal stress decreases mitotic activity, directly limiting root growth. Similarly, Bagulho et al. (2015) observed that disruption of extracellular matrix function led to 116% increased hydrogen peroxide ($H_2O_2$) production, resulting in a 77% reduction in root growth in wheat under 200 μM cadmium stress. Further evidence indicates that cadmium toxicity interferes with cell cycle regulation, particularly affecting the signaling transduction, possibly through ROS-mediated mechanisms, regardless of cyclin-dependent kinase (CDK) decline (Rehman et al., 2018). Likewise, copper (Cu) excess accumulation alters auxin distribution, resulting in the poor proliferation of primary roots (Rather et al., 2020). Likewise, copper toxicity disrupted auxin levels in Arabidopsis, bringing about root inhibition and obstructing nitric oxide functioning. On the other hand, Satbhai et al. (2015) also demonstrated that root morphology undergoes modification under such circumstances, as suppression of root elongation is often accompanied by an increase in root diameter, implying that the cytoskeleton may act as a target for metal toxicity.

Heavy metal contamination not only hampers root growth but also impairs its function of water and nutrient uptake; the plant is weak with low biomass accumulation and shoot growth (Sun et al., 2019). To counter these detrimental effects, the plant uses several defense mechanisms, such as limiting metal uptake, regulating metal transport through specific transporters, activating antioxidant systems, isolating metals in vacuoles, and developing and depositing callose in cell walls (Riyazuddin et al., 2021). These coping mechanisms and strategies, along with enhancing root adaptability, also restrict the entrance of heavy metals in root tissue. Additionally, Raven (2022), besides blocking the entrance of heavy metals, some of it is transported to the upper parts of the plant where they are segregated in vacuoles and rendered inactive. According to Yu et al. (2024), tomato seedlings exposed to nickel stress showed adaptive responses such as the synthesis of secondary metabolites, increased proline accumulation, maintenance of redox balance, regulation of ion homeostasis, and enhanced nickel uptake.

*Table 2. Effect of metal toxicity on plant growth, yield, and photosynthetic pigments different heavy metals, their resources, and occurrence forms in the soil*

| Crop name | Effect of metal toxicity | Reference |
|---|---|---|
| Asian rice | Poor seed emergence | Wang & Yu (2022) |
| Wheat | Low fresh and dry biomass of plants | Zhang et al. (2012) |
| Maize | Shortened roots and poor anchorage | Piršelová et al. (2022) |
| Tomato | Degradation of photosynthetic pigments | Rady and Osman (2012) |
| Peppers | Wilting and low turgor pressure | El-Saadony et al. (2024) |
| Eggplant | Disruptions in water relations | Alamer and Galal (2022) |
| Mango | Nutritional imbalances and deficit conditions | Riskuwa-Shehu et al. (2020) |
| Guava and mango | Higher lipid peroxidation and membrane leakage | Krishnani et al. (2021) |

Disruption of plant metabolism and reduced biomass accumulation are also linked to heavy metal toxicity (Table 2). New leaves, when exposed to heavy metal, show chlorosis earlier than older leaves, along with shorter stem and root lengths. At the biochemical level, heavy metals can destroy and deactivate enzymes, inhibit essential metabolic pathways, cause hormonal disorders, upset protein synthesis, interfere with DNA repair, and negatively affect seed germination, color formation, photosynthetic activity, gas exchange, and plant respiration (Nagajyoti et al., 2010). Furthermore, Laude and Prior (2004) observed that at the cellular level, the plasma membranes act as strict regulatory barriers that regulate the entrance of any toxic substance into the cell, saving the plant from stress-induced damage. Proving a similar point, Schmid et al. (2005) claimed that the heavy metal tolerance of *Arabidopsis halleri* and *Arabidopsis arenosa* was superior to that of *Arabidopsis thaliana* as their membranes depolarized at slower paces under stress, indicating that inspecting potential changes in membranes can provide useful insights into plant responses to metal toxicity.

Sassafras is an aromatic deciduous tree. Its seedlings exhibited a significant reduction in photosynthetic rate and intracellular $CO_2$ concentration under cadmium toxicity in the research carried out by Farid et al. (2013). A similar study carried out by Cabrita et al. (2018) demonstrated that metals such as cobalt, copper, chromium, lead, and mercury diminished chlorophyll content and lowered the efficiency of photochemical capacity of photosystem II (PS II) in *Phaeodactylum tricornutum*. Similarly, Emamverdian and Ding (2017) demonstrated similar effects of zinc, copper, cadmium, and lead on durum wheat, as their exposure resulted in lower PS II quantum yield as well as chlorophyll and carotenoid content, underscoring the adverse effects of heavy metals on photosynthetic processes. Another study by Khodanitska et al. (2021) also demonstrated the detrimental effects of heavy metals on the crucial function of nitrogen metabolism, compromising basic plant activities such as disbursement of nutrients for plant growth. Heavy metals suppress nitrate-related enzymatic activities by increasing protease activity. Cadmium exposure inhibits key enzymes such as nitrate reductase and glutamine synthetase (GS), thereby disrupting nitrogen absorption, uptake, and utilization in plants, as shown by Singh et al. (2019).

## HEIGHTENING OF OXIDATIVE STRESS IN PLANTS DUE TO METAL TOXICITY

Heavy metals (HMs) and metalloids such as arsenic, cadmium, lead, chromium, aluminum, mercury, and nickel, when they exceed safe limits, induce adverse effects on physiological mechanisms and disrupt crucial metabolic functions of plant development. The studies carried out by Steffens et al. (2014) and Xu et al. (2018) indicate that chromium toxicity promotes ROS generation, leading to oxidative damage.

Heavy metals up to the threshold levels interfere with normal metabolic activities through different mechanisms, by disrupting and displacing protein synthesis blocks (Hall et al., 2002). This also undermines the efficacy and integrity of the cytoplasmic membrane, impairing basic functions like respiration, photosynthesis, and enzymatic activity (Hossain et al., 2012). Exposure to metal toxicity initiates a range of complex signaling pathways in plants, which includes the activation of mitogen-activated protein kinases (MAPKs), which play a crucial role in signal transduction by regulating gene expression in the nucleus in relation to environmental changes. Additionally, calcium-dependent signaling, hormonal regulation, and ROS-mediated signaling pathways are also involved in the plant's stress response (Kumar et al., 2016).

Heavy metal toxicity can cause alteration of plant proteins, induce osmotic and oxidative stress, which causes plants to initiate defensive mechanisms and adaptive cellular responses such as accelerated antioxidant defense system, hastening ROS scavenging, synthesis of stress-related proteins, and accumulation of compatible solutes (Zhu et al., 2002). Several solutes are manufactured by plants under stress, among them the most prevalent are sucrose, proline, polyols, trehalose, and quaternary ammonium compounds, including glycine betaine, hydroxyproline betaine, alanine betaine, pipecolate betaine, and choline-O-sulfate. These compounds are also referred to as osmoprotectants as they protect plants from water-stress damage. They enhance stress resistance by stabilizing proteins and enzymes, safeguarding the membrane, and aiding in the ROS detoxification (Ashraf et al., 2007).

## ACTIVATION OF ANTIOXIDANT ASSAYS TO COPE WITH THE ADVERSE EFFECTS OF METAL TOXICITY IN PLANTS

The expression "antioxidant" refers to such compounds that can neutralize ROS without being converted into harmful radicals themselves. The efficacy of the antioxidant defense system relies upon various aspects, including the compartmentalization of ROS, the synthesis, transport, and localization of antioxidants. It also depends on the plant's capacity and strength to activate defense mechanisms and coordinate interactions among different antioxidant components (Blokhina et al., 2003).

The prime examples of ROS-detoxifying enzymes are superoxide dismutase (SOD), catalase (CAT), peroxidases (POD), and glutathione reductase (GR). SOD exists in multiple isoforms, which are mitochondrial Mn-SOD, chloroplastic Fe-SOD, and paroxysmal Cu/Zn-SOD found in the cytosol, chloroplasts, or peroxisomes. These enzymes catalyze the conversion of superoxide radicals into hydrogen peroxide ($H_2O_2$) and molecular oxygen, thereby maintaining ROS at balanced levels (Mittova et al., 2004).

CAT can break down $H_2O_2$ into water and oxygen within peroxisomes, and it does so without requiring an additional substrate. In other cellular components, namely cytosol, vacuole, cell wall, and extracellular space, $H_2O_2$ is detoxified by various peroxidases such as guaiacol peroxidase (GPX) and ascorbate peroxidase (APX, EC), which rely upon particular substrates for their activity (Noctor and Foyer, 1998; Ahmad et al., 2024; Song et al., 2026).

Under normal physiological conditions and metabolic processes, all living organisms generate ROS at an optimum level. These include superoxide anions ($O_2^-$), hydroxyl radicals (•OH), and hydrogen peroxide ($H_2O_2$). According to Ansari et al. (2012), these reactive molecules are not efficiently and swiftly removed from the plant cells; they can damage several essential biomolecules and may ultimately lead to cell death. To prevent such damage, cells rely on antioxidant systems, including molecules like ascorbate and glutathione, along with enzymes such as SOD, CAT, and POD.

In eukaryotic cells, mitochondria serve as the key site for ROS production. It was observed by Bashir et al. (2007) that superoxide radicals are formed during respiration, mainly due to the auto-oxidation of components of the mitochondrial electron transport chain. These radicals are quickly converted into $H_2O_2$ by SOD. In the findings of Creissen et al. (1994), it has been stated that ROS are commonly produced as a result of metabolic activities in plants, but under stress, their production is accelerated and is also produced in other organelles such as chloroplasts, peroxisomes, the endoplasmic reticulum, and plasma membranes.

The complexity of ROS production sites is directly related to the complexity of antioxidant defense systems. Detoxification of ROS is usually carried out in a coordinated pattern by the enzymatic antioxidants. For instance, SOD converts superoxide radicals into $H_2O_2$, which must then be further detoxified to prevent oxidative damage or the formation of highly reactive hydroxyl radicals in the presence of metal ions (Ansari et al., 2012).

$H_2O_2$ can be scavenged through various enzymes, including catalase, peroxidases, ascorbate peroxidase, glutathione peroxidase, and thioredoxin peroxidase. When ascorbate peroxidase is utilized, ascorbate serves as a primary agent for neutralizing peroxides. Following this reaction, the oxidized form of ascorbate is replenished through the enzymatic activity of monodehydroascorbate reductase (MDHAR) (Arora et al., 2002). Similarly, glutathione peroxidase and glutathione reductase operate in coordination to maintain a continuous detoxification cycle. Additionally, oxidases can provide alternative electron pathways, thereby reducing ROS generation under conditions where the electron transport chain is impaired (Altaf et al., 2024).

SOD enzymes play a central role in antioxidant defense by catalyzing the dismutation of superoxide radicals into oxygen and hydrogen peroxide (Altaf et al., 2023). This reaction involves the cycling of a metal cofactor between different oxidation states, enabling continuous detoxification (Ansari et al., 2012). CAT is another vital antioxidant enzyme that is widely present in living organisms and facilitates the rapid decomposition of $H_2O_2$ into water and oxygen (Chelikani et al., 2004). Due to its exceptionally high turnover rate, a single CAT molecule can convert millions of $H_2O_2$ molecules per second, highlighting its critical role in protecting cells from oxidative damage (Morana et al., 2012).

## EFFECT OF METAL TOXICITY ON SIGNALING MOLECULES AND ENDOGENOUS HORMONES

Contamination of agricultural soils caused by heavy metals such as arsenic, lead, nickel, copper, cadmium, zinc, and chromium has been reported all over the world, with contamination severity ranging from low to severe (Rasheed et al., 2021). Likewise, Hussain et al. (2021) and Yasir et al. (2021) demonstrated the sources of these hazardous pollutants and stated that these originate from multiple anthropogenic sources, including prolonged use of phosphate-based fertilizers, usage of sewage sludge, emissions from smelting activities, industrial and medical waste disposal, and inefficient irrigation practices.

Exposure to elevated levels of heavy metals triggers excessive production of reactive oxygen species (ROS) in plants, which severely compromises their growth, development, and survival (Rahman et al., 2022). Plants mitigate such stress through a series of coordinated defense responses that include perception of external stress signals, transmission and processing of these signals within the cell, and activation and mobilization of adaptive mechanisms that modify physiological, biochemical, and molecular processes to minimize stress-induced damage (Patel et al., 2021). Dubey et al. (2021) also reported that metal contamination enhances protease activity while simultaneously reducing the activity of key enzymes involved in nitrogen metabolism, including nitrite reductase (NiR), nitrate reductase (NR), glutamate dehydrogenase, and glutamine synthetase. Therefore, this disruption can adversely affect nitrogen assimilation and overall plant metabolism (Muhammad et al., 2024; El-Mogy et al., 2025).

Hormonal imbalance is caused by heavy metal stress within plants, which has a crucial role in regulating stress responses (Muhammad et al., 2024). Plant hormones, particularly auxins such as indole-3-acetic acid (IAA), indole-3-butyric acid (IBA), and naphthalene acetic acid (NAA), are vital for maintaining proper growth and coordinating responses to metalloid-induced stress when present in optimal concentrations (Naz et al., 2022). Arsenic (As) toxicity has been shown to significantly disrupt auxin levels, resulting in impaired functioning of plant regulatory and adaptive mechanisms (Manzoor et al., 2025).

## CONCLUSION

Metal toxicity in soils and plants arises from both anthropogenic and natural activities, including industrial events, agrochemical practices, irrigation with contaminated water, and soil physicochemical factors such as pH, which affect metal bioavailability and uptake. Metal toxicity in plants results in stunted growth, reduced yield, and irregular developmental processes. The mitigation of metal toxicity is in need of time. However, modulation of physiological and biochemical indices encourages improving the plant defense system against metal toxicity. However, different management practices, such as organic amendments, phytohormones, and balanced fertilization, can be effective for the mitigation of metal toxicity in plants.

## REFERENCES


Ahmad, R., Muhammad, H. M. D., Naz, S., Tiwari, R. K., Lal, M. K., Ahmad, P., & Altaf, M. A. (2024). From stress to success: Silicon and nano-silicon strategies for enhancing fruit yield and quality in cadmium-exposed trees. *South African Journal of Botany*, *167*, 441–447. DOI: 10.1016/j.sajb.2024.02.042

Alamer, K. H., & Galal, T. M. (2022). Safety assessment and sustainability of consuming eggplant (*Solanum melongena* L.) grown in wastewater-contaminated agricultural soils. *Scientific Reports*, *12*(1), 9768. DOI: 10.1038/s41598-022-13992-7 PMID: 35697742

Altaf, M. A., Naz, S., Kumar, R., Sardar, H., Nawaz, M. A., Kumar, A., Lal, P., Ahmad, R., Hayat, F., Wani, M. A., Tiwari, R. K., & Lal, M. K. (2023). Unraveling the mechanisms of cadmium toxicity in horticultural plants: Implications for plant health. *South African Journal of Botany*, *163*, 433–442. DOI: 10.1016/j.sajb.2023.10.064

Altaf, M. A., Shahid, R., Naz, S., Ahmad, R., Manzoor, M. A., Alsahli, A. A., Altaf, M. M., & Ahmad, P. (2024). Vanadium toxicity was alleviated by supplementation of silicon in tomato seedlings: Upregulating antioxidative enzymes and glyoxalase system. *Plant Physiology and Biochemistry : PPB*, *213*, 108809. DOI: 10.1016/j.plaphy.2024.108809 PMID: 38875778

Ansari, M. K. A., Anjum, N. A., Ahmad, A., Umar, S., & Iqbal, M. (2012). Heavy metals in soil and plants: An overview of arsenic, cadmium, chromium and mercury. In N. A. Anjum, S. Umar, & A. Ahmad (Eds.), *Oxidative stress in plants: Causes, consequences and tolerance* (pp. 499–518). I.K. International Publishing House.

Arora, A., Sairam, R. K., & Srivastava, G. C. (2002). Oxidative stress and antioxidative system in plants. *Current Science*, *82*(10), 1227–1238.

Ashraf, M., & Foolad, M. R. (2007). Roles of glycine betaine and proline in improving plant abiotic stress resistance. *Environmental and Experimental Botany*, *59*(2), 206–216. DOI: 10.1016/j.envexpbot.2005.12.006

Bagulho, A., Vilas-Boas, F., Pena, A., Peneda, C., Santos, F. C., Jerónimo, A., de Almeida, R. F., & Real, C. (2015). The extracellular matrix modulates $H_2O_2$ degradation and redox signaling in endothelial cells. *Redox Biology*, *6*, 454–460. DOI: 10.1016/j.redox.2015.09.006 PMID: 26409032

Basharat, M. (2019). Water management in the Indus Basin in Pakistan: Challenges and opportunities. In *Indus River Basin* (pp. 375–388). Elsevier. DOI: 10.1016/B978-0-12-812782-7.00017-5

Bashir, F., Mahmooduzzafar, , Siddiqi, T. O., & Iqbal, M. (2007). The antioxidative response system in *Glycine max* (L.) Merr. exposed to deltamethrin, a synthetic pyrethroid insecticide. *Environmental Pollution*, *147*(1), 94–100. DOI: 10.1016/j.envpol.2006.08.013 PMID: 17055627

Bijekar, S., Padariya, H. D., Yadav, V. K., Gacem, A., Hasan, M. A., Awwad, N. S., Yadav, K. K., Islam, S., Park, S., & Jeon, B. H. (2022). The state of the art and emerging trends in wastewater treatment in developing nations. *Water (Basel)*, *14*(16), 2537. DOI: 10.3390/w14162537

Blokhina, O. B., Virolainen, E., & Fagerstedt, K. V. (2003). Antioxidants, oxidative damage and oxygen deprivation stress: A review. *Annals of Botany*, *91*(2), 179–194. DOI: 10.1093/aob/mcf118 PMID: 12509339

Bystrická, J., Árvay, J., Musilová, J., Vollmannová, A., Tóth, T., & Lenková, M. (2016). The investigation of sensitivity of different types of onion to heavy metal intake from contaminated soil. *International Journal of Environmental Research*, *10*(3), 427–440.

Cabrita, M. T., Duarte, B., Gameiro, C., Godinho, R. M., & Caçador, I. (2018). Photochemical features and trace element substituted chlorophylls as early detection biomarkers of metal exposure in the model diatom *Phaeodactylum tricornutum*. *Ecological Indicators*, *95*, 1038–1052. DOI: 10.1016/j.ecolind.2017.07.057

Chelikani, P., Fita, I., & Loewen, P. C. (2004). Diversity of structures and properties among catalases. *Cellular and Molecular Life Sciences : CMLS*, *61*(2), 192–208. DOI: 10.1007/s00018-003-3206-5 PMID: 14745498

Creissen, G. P., Edwards, E. A., & Mullineaux, P. M. (1994). Glutathione reductase and ascorbate peroxidase. In Foyer, C. H., & Mullineaux, P. M. (Eds.), *Causes Of Photoxidative Stress and Amelioration of Defense Systems in Plants* (pp. 343–364). CRC Press.

Dickin, S. K., Schuster-Wallace, C. J., Qadir, M., & Pizzacalla, K. (2016). A review of health risks and pathways for exposure to wastewater use in agriculture. *Environmental Health Perspectives, 124*(7), 900–909. DOI: 10.1289/ehp.1509995 PMID: 26824464

Dubey, R. S., Srivastava, R. K., & Pessarakli, M. (2021). Physiological mechanisms of nitrogen absorption and assimilation in plants under stressful conditions. In Pessarakli, M. (Ed.), *Handbook of Plant and Crop Physiology* (pp. 579–616). CRC Press. DOI: 10.1201/9781003093640-36

El-Mogy, M. M., Jan, M. F., Naz, I., Ahmad, I., Ahmad, R., & Altaf, M. T. (2025). Nanoparticle innovations for mitigating metal toxicity in plants. *Phyton, 94*(3), 623–640. DOI: 10.32604/phyton.2025.063763

El-Saadony, M. T., Desoky, E. S. M., El-Tarabily, K. A., AbuQamar, S. F., & Saad, A. M. (2024). Exploiting the role of plant growth-promoting rhizobacteria in reducing heavy metal toxicity of pepper (*Capsicum annuum* L.). *Environmental Science and Pollution Research International, 31*(18), 27465–27484. DOI: 10.1007/s11356-024-32874-1 PMID: 38512572

Emamverdian, A., & Ding, Y. (2017). Heavy metals induced changes on growth, antioxidant enzyme activity, gas exchange parameters and protein structures in *Sasa kongosanensis* f. *Aureo-striatus. Polish Journal of Environmental Studies, 26*(2), 585–592. DOI: 10.15244/pjoes/65367

Engwa, G. A., Ferdinand, P. U., Nwalo, F. N., & Unachukwu, M. N. (2019). Heavy metal toxicity in humans. In *Poisoning in the modern world: New tricks for an old dog* (pp. 77–100). IntechOpen. DOI: 10.5772/intechopen.82511

Farid, M., Shakoor, M. B., Ehsan, S., Ali, S., Zubair, M., & Hanif, M. S. (2013). Morphological, physiological and biochemical responses of different plant species to Cd stress. *International Journal of Chemical and Biochemical Science, 3*, 53–60.

Grattan, S. R., Díaz, F. J., Pedrero, F., & Vivaldi, G. A. (2015). Assessing the suitability of saline wastewaters for irrigation of *Citrus* spp.: Emphasis on boron and specific-ion interactions. *Agricultural Water Management, 157*, 48–58. DOI: 10.1016/j.agwat.2015.01.002

Guarino, C., Paura, B., & Sciarrillo, R. (2018). Enhancing phytoextraction of heavy metals at real scale by combining salicaceae trees with microbial consortia. *Frontiers in Environmental Science, 6*, 137. DOI: 10.3389/fenvs.2018.00137

Hassan, A. S. (2019). Inorganic-based pesticides: A review article. *Egyptian Scientific Journal of Pesticides, 5*, 39–52.

Hassena, A. B., Zouari, M., Trabelsi, L., Khabou, W., & Zouari, N. (2018). Physiological improvements of young olive tree (*Olea europaea* L. cv. Chetoui) under short-term irrigation with treated wastewater. *Agricultural Water Management, 207*, 53–58. DOI: 10.1016/j.agwat.2018.05.024

Hossain, M. A., Piyatida, P., da Silva, J. A. T., & Fujita, M. (2012). Molecular mechanism of heavy metal toxicity and tolerance in plants: Central role of glutathione in detoxification of reactive oxygen species and methylglyoxal and in heavy metal chelation. *Journal of Botany (Egypt), 2012*, 872875. DOI: 10.1155/2012/872875

Hotamisligil, G. S., & Davis, R. J. (2016). Cell signaling and stress responses. *Cold Spring Harbor Perspectives in Biology, 8*(10), a006072. DOI: 10.1101/cshperspect.a006072 PMID: 27698029

Hussain, B., Ashraf, M. N., Abbas, A., Li, J., & Farooq, M. (2021). Cadmium stress in paddy fields: Effects of soil conditions and remediation strategies. *The Science of the Total Environment, 754*, 142188. DOI: 10.1016/j.scitotenv.2020.142188 PMID: 33254942

Jingjing, W. A. N. G., & Xiaohua, Y. U. (2022). Wastewater irrigation and crop yield: A meta-analysis. *Journal of Integrative Agriculture, 21*(4), 1215–1224. DOI: 10.1016/S2095-3119(21)63853-4

Kalaivanan, D., & Ganeshamurthy, A. N. (2016). Mechanisms of heavy metal toxicity in plants. In *Abiotic stress physiology of horticultural crops* (pp. 85–102). Springer India., DOI: 10.1007/978-81-322-2725-0_5

Khanna, P. (2011). Assessment of heavy metal contamination in different vegetables grown in and around urban areas. *Research Journal of Environmental Toxicology*, *5*(3), 162–179. DOI: 10.3923/rjet.2011.162.179

Khodanitska, O., Tkachuk, O., & Matviichuk, O. (2021). Physiological activity of plant growth stimulators. *The Scientific Heritage*, *58*(1), 36–38.

Krishnani, K. K., Choudhary, K., Boddu, V. M., Moon, D. H., & Meng, X. (2021). Heavy metals biosorption mechanism of partially delignified products derived from mango (*Mangifera indica*) and guava (*Psidium guajava*) barks. *Environmental Science and Pollution Research International*, *28*(25), 32891–32904. DOI: 10.1007/s11356-021-12874-1 PMID: 33638079

Kumar, S., & Trivedi, P. K. (2016). Heavy metal stress signalling in plants. In *Plant metal interaction: Emerging remediation techniques* (pp. 585–603). Elsevier. DOI: 10.1016/B978-0-12-803158-2.00025-4

Laude, A. J., & Prior, I. A. (2004). Plasma membrane microdomains: Organization, function and trafficking. *Molecular Membrane Biology*, *21*(3), 193–205. DOI: 10.1080/09687680410001700517 PMID: 15204627

Manzoor, M., Korolev, K., & Ahmad, R. (2025). Zinc oxide nanoparticles: Abiotic stress tolerance in fruit crops focusing on sustainable production. *Phyton*, *94*(5), 1401–1418. DOI: 10.32604/phyton.2025.063930

Mi, B., Liu, F., Xie, L., Zhou, H., Wu, F., & Dai, X. (2019). Evaluation of the uptake capacities of heavy metals in Chinese cabbage. *Ecotoxicology and Environmental Safety*, *171*, 511–517. DOI: 10.1016/j.ecoenv.2019.01.022 PMID: 30641311

Mishra, R. K. (2023). Fresh water availability and its global challenge. *British Journal of Multidisciplinary and Advanced Studies*, *4*(3), 1–78. DOI: 10.37745/bjmas.2022.0208

Mittova, V., Guy, M., Tal, M., & Volokita, M. (2004). Salinity upregulates the antioxidative system in root mitochondria and peroxisomes of the wild salt-tolerant tomato species *Lycopersicon pennellii*. *Journal of Experimental Botany*, *55*(399), 1105–1113. DOI: 10.1093/jxb/erh113 PMID: 15047761

Mourato, M. P., Moreira, I. N., Leitão, I., Pinto, F. R., Sales, J. R., & Louro Martins, L. (2015). Effect of heavy metals in plants of the genus *Brassica*. *International Journal of Molecular Sciences*, *16*(8), 17975–17998. DOI: 10.3390/ijms160817975 PMID: 26247945

Muhammad, H. M. D., Naz, S., Lal, M. K., Tiwari, R. K., Ahmad, R., Nawaz, M. A., Das, R., & Altaf, M. A. (2024). Melatonin in business with abiotic stresses in vegetable crops. *Scientia Horticulturae*, *324*, 112594. DOI: 10.1016/j.scienta.2023.112594

Nagajyoti, P. C., Lee, K. D., & Sreekanth, T. V. M. (2010). Heavy metals, occurrence and toxicity for plants: A review. *Environmental Chemistry Letters*, *8*(3), 199–216. DOI: 10.1007/s10311-010-0297-8

Naz, S., Anjum, M. A., Sadiq, B., Ahmad, R., Altaf, M. A., El-Sheikh, M. A., & Shakoor, A. (2022). Purification of sewage wastewater through sand column filter for lessening heavy metal accumulation in lettuce, carrot, and cauliflower. *Water (Basel)*, *14*(22), 3770. DOI: 10.3390/w14223770

Noctor, G., & Foyer, C. H. (1998). Ascorbate and glutathione: Keeping active oxygen under control. *Annual Review of Plant Physiology and Plant Molecular Biology*, *49*(1), 249–279. DOI: 10.1146/annurev.arplant.49.1.249 PMID: 15012235

Patel, M., Surti, M. M., Ashraf, S. A., & Adnan, M. (2021). Physiological and molecular responses to heavy metal stresses in plants. In *Harsh environment and plant resilience* (pp. 171–202). DOI: 10.1007/978-3-030-65912-7_8

Paudel, I., Bar-Tal, A., Levy, G. J., Rotbart, N., Ephrath, J. E., & Cohen, S. (2018). Treated wastewater irrigation: Soil variables and grapefruit tree performance. *Agricultural Water Management*, *204*, 126–137. DOI: 10.1016/j.agwat.2018.04.006

Piotto, F. A., Carvalho, M. E. A., Souza, L. A., Rabêlo, F. H. S., Franco, M. R., Batagin-Piotto, K. D., & Azevedo, R. A. (2018). Estimating tomato tolerance to heavy metal toxicity: Cadmium as study case. *Environmental Science and Pollution Research International*, *25*(27), 27535–27544. DOI: 10.1007/s11356-018-2778-4 PMID: 30051291

Piršelová, B., Galuščáková, Ľ., Lengyelová, L., Kubová, V., Jandová, V., & Hegrová, J. (2022). Assessment of the hormetic effect of arsenic on growth and physiology of two cultivars of maize (*Zea mays* L.). *Plants, 11*(24), 3433. DOI: 10.3390/plants11243433 PMID: 36559544

Pretty, J. (2008). Agricultural sustainability: Concepts, principles and evidence. *Philosophical Transactions of the Royal Society of London. Series B, Biological Sciences, 363*(1491), 447–465. DOI: 10.1098/rstb.2007.2163 PMID: 17652074

Qadir, M., Wichelns, D., Raschid-Sally, L., McCornick, P. G., Drechsel, P., Bahri, A., & Minhas, P. S. (2010). The challenges of wastewater irrigation in developing countries. *Agricultural Water Management, 97*(4), 561–568. DOI: 10.1016/j.agwat.2008.11.004

Rahman, M. H., Roy, B., Chowdhury, G. M., Hasan, A., & Saimun, M. S. R. (2022). Medicinal plant sources and traditional healthcare practices of forest-dependent communities in and around Chunati Wildlife Sanctuary in southeastern Bangladesh. *Environmental Sustainability, 5*(2), 207–241. DOI: 10.1007/s42398-022-00230-z PMID: 37521586

Rasheed, A., Hassan, M. U., Fahad, S., Aamer, M., Batool, M., Ilyas, M., & Li, H. (2021). Heavy metals stress and plants defense responses. In *Sustainable soil and land management and climate change* (pp. 57–82). CRC Press.

Riyazuddin, R., Nisha, N., Ejaz, B., Khan, M. I. R., Kumar, M., Ramteke, P. W., & Gupta, R. (2021). A comprehensive review on the heavy metal toxicity and sequestration in plants. *Biomolecules, 12*(1), 43. DOI: 10.3390/biom12010043 PMID: 35053191

Rroji, M., Figurek, A., Viggiano, D., Capasso, G., & Spasovski, G. (2022). Phosphate in the context of cognitive impairment and other neurological disorders occurrence in chronic kidney disease. *International Journal of Molecular Sciences, 23*(13), 7362. DOI: 10.3390/ijms23137362 PMID: 35806367

Sinha, P., Shukla, A. K., & Sharma, Y. K. (2015). Amelioration of heavy-metal toxicity in cauliflower by application of salicylic acid. *Communications in Soil Science and Plant Analysis*, *46*(10), 1309–1319. DOI: 10.1080/00103624.2015.1033543

Song, C., Bibi, Z., Ali, S., & Ahmad, R. (2026). Silicon mitigates aluminum toxicity in lettuce by improving photosynthetic pigments and antioxidant defense system. *Biometals*, 1–14. DOI: 10.1007/s10534-026-00819-5 PMID: 42050262

Sun, Z., Su, C., Yun, J., Jiang, Q., Wang, L., Wang, Y., Cao, D., Zhao, F., Zhao, Q., Zhang, M., Zhou, B., Zhang, L., Kong, F., Liu, B., Tong, Y., & Li, X. (2019). Genetic improvement of the shoot architecture and yield in soya bean plants via the manipulation of GmmiR156b. *Plant Biotechnology Journal*, *17*(1), 50–62. DOI: 10.1111/pbi.12946 PMID: 29729214

Vojodi Mehrabani, L. (2019). The effects of methanol and ethanol foliar application under salinity stress on some physiological characteristics of Pelargonium graveolens L. *Journal of Plant Physiology and Breeding*, *9*(1), 63–73.

Wang, J., & Yu, X. (2022). Wastewater irrigation and crop yield: A meta-analysis. *Journal of Integrative Agriculture*, *21*(4), 1215–1224. DOI: 10.1016/S2095-3119(21)63853-4

Woldetsadik, D., Drechsel, P., Keraita, B., Itanna, F., & Gebrekidan, H. (2018). Farmers' perceptions on irrigation water contamination, health risks and risk management measures in prominent wastewater-irrigated vegetable farming sites of Addis Ababa, Ethiopia. *Environment Systems & Decisions*, *38*(1), 52–64. DOI: 10.1007/s10669-017-9665-2

Xie, Q., Wei, X., Liu, Y., Han, F., & Li, Z. (2022). Germplasm enhancement and identification of loci conferring resistance against Plasmodiophora brassicae in broccoli. *Genes*, *13*(9), 1600. DOI: 10.3390/genes13091600 PMID: 36140766

Yap, C. K., Razali, A., Nulit, R., Peng, S. H. T., Yap, C. W., Okamura, H., & Ismail, M. S. (2021). *Biomonitoring of heavy metals in the guava (Psidium guajava) for their health risk assessment in Kluang*. Food Science and Engineering.

Yeşilyurt, S. (2023). Phytoremediation method and Brassica family: Removal of chromium, cadmium and lead accumulation with broccoli (Brassica oleracea var. italica). *Results in Chemistry*, *6*, 101005. DOI: 10.1016/j.rechem.2023.101005

Zhu, J. K. (2002). Salt and drought stress signal transduction in plants. *Annual Review of Plant Physiology and Plant Molecular Biology*, *53*(1), 247–273. DOI: 10.1146/annurev.arplant.53.091401.143329 PMID: 12221975

# Chapter 2
# Plant Growth Regulators Trigger Stress Tolerance in Plants Against Metal Toxicity

**Saqib Ali**
http://orcid.org/0000-0002-6338-4385
*The University of Agriculture, Dera Ismail Khan, Pakistan*

**Nabil Touzout**
http://orcid.org/0000-0002-7091-9990
*Laboratory Materials and Environment, University Yahia Fares of Medea, Medea, Algeria*

**Riaz Ahmad**
http://orcid.org/0000-0002-4383-8335
*The University of Agriculture, Dera Ismail Khan, Pakistan*

## ABSTRACT

*Climate change has disrupted farming practices due to abiotic stress conditions. Metal toxicity is a major threat to agriculture. The drastic growth reduction, poor yield, and inferior quality were recorded from the metal toxicity. Metal toxicity contributes to the impairment of food safety and exhibits adverse effects on crop production. Plant growth regulators*

DOI: 10.4018/979-8-3693-9899-9.ch002

*(PGRs) are well-known because of their excellent contribution to the mitigation of the adverse effects of metal toxicity occurring in plants. The antioxidant defense system of plants is boosted with the exogenous supplementation of PGRs. Different aspects, such as uptake, translocation, and detoxification of metals, can be regulated with PGRs supplementation. The harmful effects of metals on plants were significantly mitigated with exogenous PGRs supplementation. Phytohormone use is a long-term approach to reducing the adverse effects of metal stress conditions. PGRS are an excellent opportunity for the mitigation of adverse effects of metal toxicity in plants by improving growth, yield, and quality.*

## INTRODUCTION

Toxic metals in plants are an urgent ecological issue because it has several harmful effects on plant growth and development (Abou Fayssal et al., 2024). Excessive metals cause oxidative harm and impairment of cells and tissues by generating reactive oxygen species (ROS) (Dubey, 2010; Rucińiska-Sobkowiak, 2010). Reduced root growth, impaired photosynthesis, and nutritional imbalances are typical symptoms (Maleki et al., 2017). To counteract metal toxicities, plants use an array of strategies, including triggering antioxidant responses and generating metal-binding substances such as phytochelatins (Fryzova et al., 2017). Molecular implications, such as microRNAs, are also contributing to the metal tolerance mechanism in plants (Nowicka, 2022; Manzoor et al., 2025). Transcription factors and many other genes are also activated by these microRNAs, which further regulate growth inhibition. The advancements in the mitigation strategies need time to attain higher crop productivity from plants growing under polluted soils (Mansoor et al., 2023; Muhammad et al., 2024).

Exogenous supplementation of various plant growth regulators (PGRs) greatly enhanced plant tolerance against metal toxicity. These are effective in enhancing the plant defense system against metal toxicity (Chen et al., 2021). Moreover, these phytohormones are naturally produced within the plants to improve biochemical activities under stress conditions (Rostami and Azhdarpoor, 2019). The regulation of metallic ions improves metabolic activities, and activation of the defense system results in higher production of crops by reducing oxidative stress (Tassi et al., 2008). Abscisic acid (ABA) is one of the essential PGRs for metal elimination and stress regulation by altering root structure. ABA limits metal absorption and causes stomatal closure to reduce water loss. Likewise, it has been proven that salicylic acid reduces metal toxicity by boosting enzymatic activity, which together scavenge poisonous ROS (Kumar et al., 2022). Furthermore, ascorbate as well as glutathione pools are two important components of redox homeostasis that are modulated by salicylic acid (Huang et al., 2021).

Gibberellins and cytokinins, which are usually linked to growth excitement, have also been linked to reducing stress (Sytar et al., 2019). Gibberellins improve nutrient uptake and encourage cell elongation, combating the inhibition of growth imposed due to metal toxicity. Thus, cytokinin assists in preserving chlorophyll levels and also improves photosynthesis, while delaying leaf death under distress (Bulak et al., 2014). Moreover, cytokinin is involved in the activation of defense signals. Jasmonates and ethylene also aid in stress adaptation through the induction of particular genes that respond to stress conditions (Asgher et al., 2015). The improvement of metal retention into vessels, and the facilitation of chelation were also recorded through organic acids and phytochelatins. Among phytohormones, brassinosteroids (BRs) play a significant role in plant stress by changes in physiological and biochemical changes. Exogenous BRs supplementation improves endogenous BRs levels under normal and stress conditions. BRs regulate membrane integrity, activate antioxidant enzymes, and modulate expression of metal carrier genes (Kapoor et al., 2022; Jan et al., 2024).

Polyamines application illustrates a significant contribution in the mitigation of metal toxicity by altering defense-related paths, ROS elimination, and restoring cell wall functioning during metal stress (Ashfaque et al., 2020; Asija et al., 2023). These polyamines restrict metal uptake and translocation within plant cells and organelles. The expression of genes improves with polyamine application in plants growing in metal-toxic conditions (Blázquez, 2024).

PGRs are a better way to improve tolerance in plants subjected to metal toxicity (Zulfiqar and Ashraf, 2021; Mondal et al., 2022). PGRs are involved in the regulation of stress signaling, metal elimination, and growth improvements (Pardo-Hernández et al., 2021; Muhamad et al., 2025). Moreover, the productivity of crops can improve with exogenous supplementation of PGRs growing under adverse conditions (Pál et al., 2018). However, the development of metal-tolerant accessions is necessary for higher yield, but it takes time. Exogenous PGRs are a better solution to improve stress tolerance in plants. However, identification of tolerant/sensitive accessions is necessary for plant researchers and farmers focusing on higher productivity of crops (Jan et al., 2024; El-Mogy et al., 2025). Hence, the current chapter aims to explore the beneficial role of PGRs in the productivity of crops subjected to metal toxicity.

## EFFECT OF METAL TOXICITY ON PLANT MORPHO-PHYSIOLOGICAL AND BIOCHEMICAL ACTIVITIES

Food security, ecological sustainability, and agricultural production are all negatively impacted by the severity of metals, which has become an important ecological issue (Hou et al., 2020). Plant performance is seriously threatened by the increasing levels of pollutants in the soil as well as water, which are mostly caused by human activities, including mining, industrial expulsion, inappropriate fertilizer and pesticide use, and sewage solids (Rathod et al., 2024). The majority of metals are harmful even at low doses and are not essential, in contrast to critical micronutrients, which are required in trace levels (Oladimeji et al., 2024). These metals affect the basic morpho-physiological as well as biochemical processes of plants upon their get into the soil-plant mechanism, which compromises plant health as well as lowers crop yields (Nwankwo et al., 2025). Stunted growth, decreased germination, chlorosis in the leaves, browning of the roots, and aberrant organ development are physiological effects of metal exposures (Singh et al., 2020). Modifications in root structure and suppression of root growth are among the first signs of metal stress (Yang et al., 2024).

Toxic metals disrupt the overall intake of water and nutrients by interfering with cell growth as well as elongation, making roots the main location for metal buildup (Gul et al., 2021). Little fresh and dry biomass, chlorosis, low growth, leaf dropping, and smaller leaf size are major stress indications in plants (Asgher et al., 2015). The disturbances in morphophysical traits are an indication of irregular metabolic processes and lesser crop production due to the adverse effects of excessive metals. Metal toxicity disrupts several physiological functions contributing to plant growth and productivity (Khan et al., 2020). The disturbance of photosynthesis is one of the main physiological harms caused by toxic metals. The poor formation of chlorophyll content and carotenoids, higher lipid peroxidation, and impaired photosynthesis in plants under metal toxicity (Bashmakov et al., 2012). The higher oxidative harms were recorded in metal-stressed plants because of the overgeneration of ROS in plant cells. These ROS also enhanced lipid peroxidation and membrane rupturing, and leakage of ions (Ashfaque et al., 2020). Plants have their own defense process to tolerate stressful conditions. Tolerant plants exhibited a strong defense system as compared to susceptible plants (Zhang et al., 2024). The excessive metals result in oxidative stress due to alterations in physiological and biochemical indices. Metabolic disruptions and oxidative stress are disturbing overall plant health growing under polluted soils (Emamverdian et al., 2015). Hence, its mitigation is necessary for higher yield (Asif et al., 2022).

Metals in excess levels alter various metabolic processes and disturb the assimilation of nutrients. Nutrient deficiencies result from toxic metals' frequent competition for the displacement of crucial mineral ions (Witkowska et al., 2021). These can also disrupt enzyme activity by attaching to sulfhydryl groups and removing metal substrates from enzyme active sites. These reactions disrupt processes such as the synthesis of proteins, nitrogen breakdown, and combustion, as well as enzymatic activities (Ilieva and Sainova, 2023). Furthermore, within the framework of their defense reaction, plants build up stress-related mediators such as proline and soluble carbohydrates. The kind and metal concentration, soil characteristics, kind of plant, and phase of growth are some of the variables that affect the degree of toxicology of metals and the plant's responsiveness (Johnson et al., 2011; Altaf et al., 2024a). Some species of plants are appropriate for phytoremediation, a green way of cleaning contaminated soils, because they possess the capacity to hyperaccumulate and demonstrate natural resilience (Angulo-Bejarano et al., 2021). Nevertheless, metal toxicity causes serious physico-chemical problems in the majority of crop plants, which restricts their ability to flourish and produce (Abou Seeda et al.,

2022). Plant physico-chemical processes are severely hampered by metal toxicity, which presents a major obstacle to sustainable development and agricultural output (Roychowdhury et al., 2019). A thorough understanding of the way that plants react to metal stress is essential for the development of efficient mitigation techniques such as soil amendments, metallic material chelators, microbial restoration, and genetic modification of metal-tolerant cultivars (Baker et al., 2020; Ahmad et al., 2024).

## MITIGATION OF METAL TOXICITY WITH PGRS

Metal exposure is a serious threat to plant health and environmental pollution globally. The excessive concentrations of metals may accumulate in edible plants. The consumption of such polluted produce is a threat to human health. Toxic metals are building up in soils due to quick industrialization, mineral extraction, inappropriate waste management, and overuse of agrochemicals, all of which have a detrimental effect on the health of plants (Singhal et al., 2023). The excess of metal damages the plant cells, depresses nutrients, lessens photosynthesis, and ultimately leads to plant death by alteration of physiological and biochemical functions (Farman et al., 2019). PGRs are an effective way to improve metal tolerance by lessening adverse metal effects. The irregular physiological changes can be regulated with the application of plants under metal toxicity (Tassi et al., 2008; Altaf et al., 2024b). The antioxidant defense system plays a significant role in tolerance by improving gene expression in metal-stressed plants (Song et al., 2026). Hence, exogenous PGRs supplementation is efficient to enhance metal tolerance in plants by boosting antioxidant enzymes. Moreover, PRGs reduced oxidative injury by restoring membrane integrity (Massoud et al., 2019).

The growth and elongation of roots depend heavily on auxins. Auxin regulation is frequently upset under metal toxicity, which results in shortened roots and inadequate food intake (Zaid et al., 2020). However, extrinsic auxin use has been demonstrated to improve nutrient absorption, maintain normal root structure, and decrease metal transfer to aerial portions via altering the excretions of roots and the carrier activity (Nguyen et al., 2021). Cytokinins had the potential to reduce the adverse effects of metal toxicity by restoring chlorophyll content and carotenoids. The regular photosynthetic system improved plant biomass and overall growth traits. The reduction of oxidative harm was also recorded with exogenous cytokinins application on metal-stressed plants (Piotrowska-Niczyporuk et al., 2018; Prasad, 2022).

Gibberellins improve plant growth under metal toxicity. Moreover, the plant defense system improves with the activation of antioxidant enzymes. The higher germination rate and greater biomass can be obtained from the stress plant with the exogenous gibberellins. The low concentration of gibberellins seems sufficient for proper plant growth by reducing oxidative harm under excessive metal conditions. Plant tolerance can be improved with gibberellin supplementation against excessive metal conditions (Ogugua et al., 2022; Ullah et al., 2025). Moreover, decontamination of polluted soils is also necessary for sustainable farming practices globally.

ABA contributes to the regulation of stomatal conductance, triggering stress-responsive genes, and boosting the antioxidant system, as reported by Dar et al. (2017). ABA treatment can maintain crop metabolism under metal toxicity. The impairment of photosynthesis reduced the crop productivity (Aslam et al., 2022). Hence, ABA has a significant contribution in the mitigation of metal toxicity in plants by boosting antioxidant activities and declining oxidative injury in plants (Ahmad et al., 2024; Muhammad et al., 2024).

Salicylic acid and jasmonic acid are two important signaling phytohormones that regulate the plant defense system and improve resilience in plants against metal toxicity (Khan et al., 2012). PGRs are involved in the contribution of reducing metal toxicity by strengthening the antioxidant defense mechanism, chelating harmful substances, and triggering stress-tolerant genes (Sharma et al., 2020). Salicylic acid exhibited a significant increase in tolerance by alleviating membranes and regulating ROS-scavengers in plants against metal toxicity (Wani et al., 2017; Naz et al., 2022; Altaf et al., 2023).

BRs improve antioxidant enzymes, photosynthesis, stomatal conductance, and gene expression in metal-stressed plants. Moreover, the decline in lipid peroxidation, toxic elements, and free radicals was also recorded with BRs supplementation on metal-stressed plants (Altaf et al., 2023).

Earlier findings of Rajewska et al. (2016) illustrated that BRs had excellent properties to scavenge toxic ROS and lessen lipid peroxidation. Moreover, the regulation of signaling molecules (proline and glycine betaine) was also revealed with BRs supplementation. Hence, it has been assumed that BRs can boost the antioxidant defense system of plants subjected to metal toxicity by modifications in physiological and biochemical activities (Rajewska et al., 2016; Muhammed et al., 2025).

Jasmonates use improves tolerance in metal-stressed plants by the production of proteins that help in the reduction of metal toxicity (Wang et al., 2021). Methyl jasmonates (MeJA) exhibited excellent potential in the mitigation of cadmium toxicity in plants by improving growth performance, yield, and biochemical system. Moreover, MeJA contributes to the regulation of phytohormonal interplay in response to metal toxicity (Keramat et al., 2010; Yan et al., 2015). Rice production was enhanced with MeJA growing under arsenic toxicity. MeJA is significantly involved in the mitigation of arsenic toxicity. The decontamination of polluted soils is necessary for a suitable farming system. However, exogenous phytohormones improved the plant defense system against metal toxicity by lessening oxidative stress (Mousavi et al., 2020). The regulation of transcriptional factors was also reported under MeJA supplementation, which encourages the arsenic tolerance in rice plants. Hence, it has been revealed that JAs are excellent phytohormones that improve metal tolerance in plants (Wang et al., 2021; El-Mogy et al., 2025). Likewise, polyamines such as putrescine and spermidine can alleviate cell membrane damage, control ion movement, and boost gene expression relevant to metal toxicity (Kour et al., 2021; Manzoor et al., 2025).

ABA modulates plant growth and various developmental systems, i.e., closure of stomatal pores, embryo development, germination induction, and stress-related activities. Many crops exhibited increased endogenous ABA levels when subjected to metal toxicity (Hu et al., 2020). ABA makes an important contribution to coping with the negative impacts of metal toxicity (Kumar et al., 2020). ABA application demonstrated the reduction of cadmium toxicity in plants by inhibition of metal uptake and translocation in excess from the root towards the shoots (Lu et al., 2020). Moreover, a greater endogenous ABA level lessened the negative effects of cadmium toxicity in plants. The restriction of excessive concentrations of metals in the root zone is an excellent way of mitigating metal toxicity in plants (Hsu and Kao, 2003; Yu et al., 2023; Zhao et al., 2023). Exogenous ABA-treated plants had the potential to cope with the adverse impact of metal toxicity on plant growth and overall productivity, as illustrated by Kumar et al. (2022). Its application controls water relations, stomatal flow, and photosynthesis (Bano et al., 2023). ABA could lessen the cadmium buildup in plant organelles with improved plant biomass, better root morphology, and better plant size and higher productivity (Tang et al., 2020). Zinc toxicity in plants was regulated with ABA treatment. Moreover, detoxification and zinc absorption were regulated with transcription factors expressed with ABA application. Moreover, the elimination of lead content in the root zone was also improved with ABA treatment. Moreover, ABA-treated plants showed better growth and higher yield as compared to untreated plants growing under metal-contaminated soils (Pan et al., 2019).

The application of ethylene is most common in the ripening of fruits. However, some researchers also reported its significant contribution in the mitigation of metal toxicity in plants (Keunen et al., 2016). Its major contribution was exhibited in the elimination of toxic ROS, which lessens lipid peroxidation. Moreover, the decline in oxidative stress was also revealed in ethylene-treated plants than untreated plants under metal toxicity (Thao et al., 2015). Ethylene interacts with the ROS channel of plants by controlling the activity of enzymes that both produce and scavenge ROS in response to metal toxicity (Steffens, 2014). Ethylene has been shown to promote the generation of ROS through triggering nicotinamide adenine dinucleotide phosphate (NADPH) oxidases amid HM toxicities (Arteca and Arteca, 2007). Ethylene-insensitive mutant Arabidopsis showed less $H_2O_2$ buildup than normal plants during mercury stress (Dar et al., 2023).

*Figure 1. Plant growth regulators improve antioxidant enzyme activities of plants under metal toxicity*

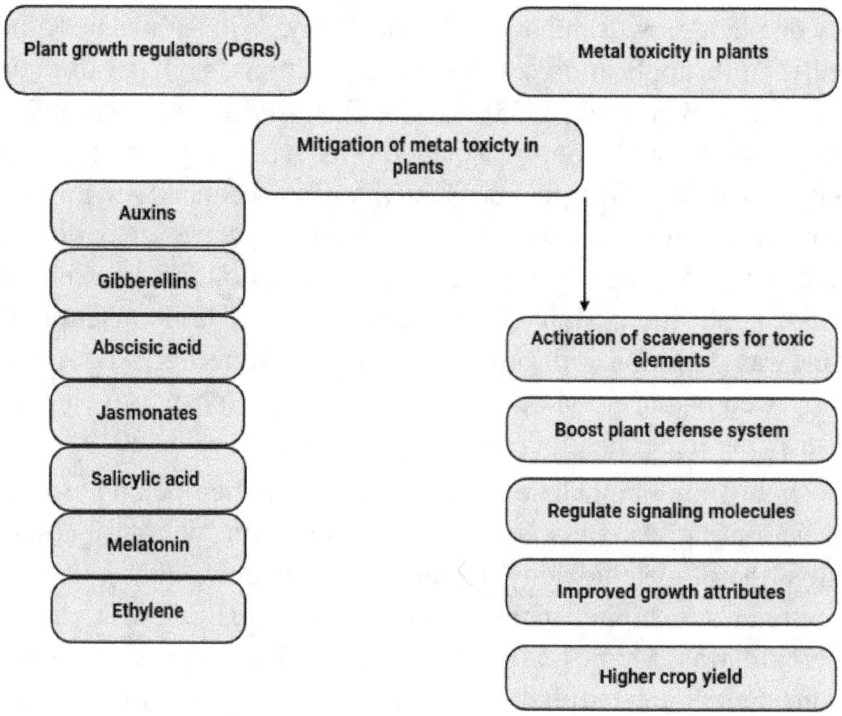


Plant metabolic and biochemical activities, including reactions to metal toxicity, are regulated by jasmonates, which are phytohormones that function as signaling molecules (Sirhindi et al., 2016). Methyl jasmonates (MeJA) have been proposed to largely increase the antioxidant defense system in plants, hence reducing the impacts of metal stress (Singh and Shah, 2014). Several enzymes that participate in ROS elimination had their enzymatic activities increased by MeJA activation. MeJA treatment under Stress greatly decreased the amounts of MDA and arsenic in leaves and roots of mustard while also increasing biomass output and fluorescent chlorophyll (Per et al., 2016). Similarly, JA improved osmolyte contents and antioxidant enzyme functioning in faba bean plants, which lessened oxidative stress by preventing the buildup of toxic substances (Manzoor et al., 2022). PGRs supplementation is a practical and environmentally responsible way to reduce the toxicity of heavy metals to plants. PGRs mitigate the negative impacts of toxic metals and encourage resilient plant growth as well as productivity in polluted soils by enhancing biochemical adaptations (Figure 1). It is essential to comprehend the distinct functions and cooperative relationships of these regulators amid metal toxicity circumstances for the purpose of establishing integrated methods for crop enhancement and environmental rehabilitation (Rostami and Azhdarpoor, 2019). The impact of different commonly used PGRs in alleviating metal toxicity (Table 1).

*Table 1. Impact of different commonly used plant growth regulators in alleviation of metal toxicity*

| **Plant growth regulators** | **Stress-relief occupation** | **Mode of action** | **References** |
|---|---|---|---|
| Abscisic acid | Boosts the antioxidant defense system | Improves enzymatic activities | Kumat et al. (2022) |
| Salicylic acid | Lessens metal uptake and translocation | Reduction in oxidative harms | Sharma et al. (2020) |
| Jasmonic acid | Encourages reclamation of toxic substances | Moderates glutathione and phytochelatin biosynthesis | Sirhindi et al. (2016) |
| Ethylene | Modulates stress signaling paths | Interrelates with signaling molecules and metal transporter genes | Steffens (2014) |
| Gibberellic acid | Encourages growth under stress | Sustains cell elongation and photosynthetic action | Ogugua et al. (2022) |
| Brassinosteroids | Progresses in membrane firmness | Induces metal chelation and antioxidant gene countenance | Gao et al. (2024) |
| Melatonin | Improves growth by regulating photosynthesis | Boosts the plant defense system in stressed plants | Pardo-Hernández et al. (2021) |

## HORMONAL CROSS-TALK TO IMPROVE METAL STRESS TOLERANCE IN PLANTS

Numerous abiotic stressors are commonly encountered by plants, and metal toxicity is one of these that is causing increasing worry because of its detrimental impact on plant growth and overall performance (Naz et al., 2022). Excess metals can seriously alter physiology and metabolism in plants (Dubey et al., 2018). These might include growth and development limitations, oxidative harm, nutrient imbalances, and reduced photosynthesis (Abou Fayssal et al., 2024). Phytohormones tightly govern an intricate network of defense responses that plants engage in response to environmental stress (Rahman et al., 2023). Phytohormones control many different kinds of physiological functions as well as stress reactions to signaling compounds (Nguyen et al., 2021). ABA triggers signaling molecules to cope with the adverse effects of metal toxicity in plants. Moreover, it also contributes to the enhancement of endogenous phytohormones, which contribute to proper plant growth, regular photosynthesis and stomatal conductance, activate antioxidant enzymes, and regulate gene transcription, and formation of chlorophyll and carotenoid contents (Yu et al., 2023). The detoxification and absorption of metals were controlled with exogenous ethylene applied to stressed plants, while SA and JAs activate defense-related genes and modulate antioxidant enzymes (Khan et al., 2012).

PGRs determine the tolerance level of plants subjected to adverse conditions. PGRs application concentration and time may vary depending upon the kind of stress, level of stress, and kind of cultivars (Alamer and Fayez, 2020). Hormonal crosstalk made a significant contribution to the mitigation of metal toxicity. Moreover, the impact of PGRs can be antagonistic and synergistic depending upon the concentrations and type of stress (Checker et al., 2018). During cadmium toxicity, the combination of ABA and JAs significantly improved the antioxidant defense system of stressed plants (Askari-Khorasgani and Pessarakli, 2019). Hence, ABA can lessen oxidative injury caused by metal toxicity by regulating ABA and JA cascades (Gao et al., 2024).

*Figure 2. Mitigation of metal toxicity with commonly used plant growth regulators*

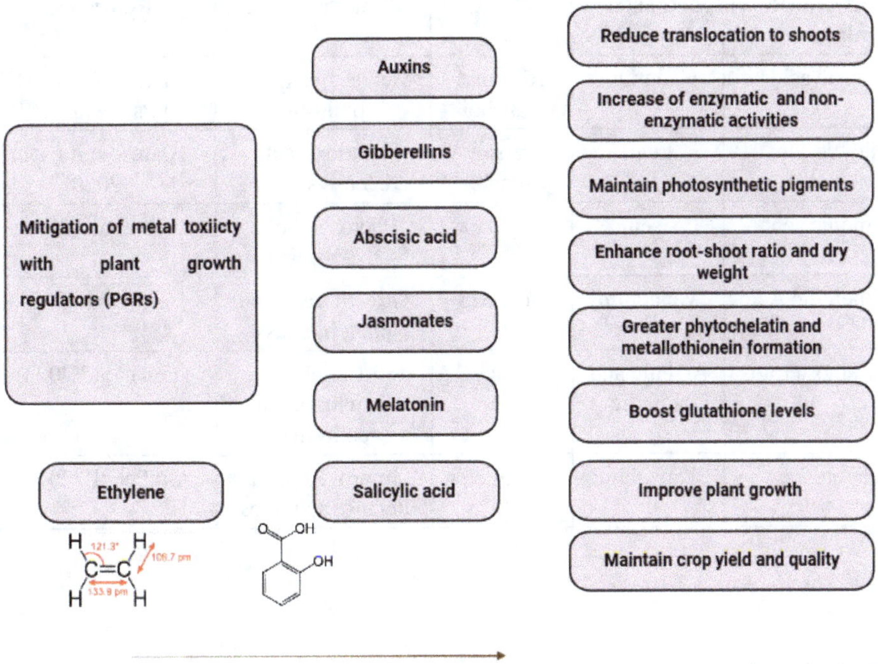


Hormonal cross-talk improves metal tolerance in plants. Exogenous PGRs effectively regulate endogenous phytohormones in plants under stress conditions (Figure 2). As a result, deciphering the intricate relationships between hormones not only improves our comprehension of the physiological processes of plant stress but also provides useful tactics for sustainable farming in areas contaminated by metals (Peñarrubia et al., 2015; Bücker-Neto et al., 2017). Effect of plant growth regulators on antioxidant enzyme activities under metal toxicity (Table 2).

*Table 2. Effect of plant growth regulators on antioxidant enzyme activities under metal toxicity*

| Plant growth regulators | Metal stress types | Crops | Traits improved | References |
|---|---|---|---|---|
| Abscisic acid | Cadmium | Rice and Pakchoi | Metals uptake and build-up | Hsu and Kao (2003) and Pan et al. (2019) |
| Salicylic acid | Lead | Parsley plants | Antioxidant defense system | Alamer and Fayez (2020) |
| Jasmonic acid | Arsenic | Eggplant | Photosynthesis mechanism | Yan et al. (2015) |
| Gibberellic Acid | Vanadium | Soybean | Growth traits and plant biomass | Basit et al. (2023) |
| Brassinosteroids | Arsenic and cadmium | Rice | Cellular detoxification pathways | Lan et al. (2023) |
| Melatonin | Vanadium | Muskmelon | Improved plant defense activities | Altaf et al. (2024b) |

## CONCLUSION

Plant growth regulators (PGRs) are essential for reducing toxic metals in plants. Metal toxicity can be eliminated by enhancing osmolyte build-up, the antioxidant system, and controlling various physiological and biochemical functions. PGRs comprising auxins, gibberellins, ethylene, brassinosteroids, jasmonic acid, and abscisic acid assist plants in overcoming elevated metal ion concentrations in plants by strengthening stress tolerance systems. By promoting antioxidant defense mechanisms, regulating the generation of toxic substances, and preserving nutritional homeostasis, these regulators also help to prevent pesticide-induced toxicity. PGRs also lessen the negative effects of herbicide-induced distress by encouraging growth, transpiration, and fundamental plant growth under adverse conditions caused by metals. PGRs applied exogenously have demonstrated potential in increasing plants' tolerance against metal toxicity. PGRs can lessen the toxicity of metals to plants. Exogenous phytohormonal management can increase plant tolerance by controlling signaling pathways linked to heavy metal tolerance.

## REFERENCES


Abou Fayssal, S., Kumar, P., Popescu, S. M., Sardar, H., Ahmad, R., Gupta, D., Gaur, S. K., Alharby, H. F., & Al-Ghamdi, A. G. (2024). Health risk assessment of heavy metals in saffron (*Crocus sativus* L.) cultivated in domestic wastewater and lake water irrigated soils. *Heliyon, 10*(5), e27138. DOI: 10.1016/j.heliyon.2024.e27138

Abou Seeda, M., Abou El-Nour, E. Z., Abdallah, M., & El-Bassiouny, H. (2022). Impacts of metal, metalloid and their effects in plant physiology: A review. *Middle East J Agric Res*, *11*(3), 838–931.

Ahmad, R., Muhammad, H. M. D., Naz, S., Tiwari, R. K., Lal, M. K., Ahmad, P., & Altaf, M. A. (2024). From stress to success: Silicon and nano-silicon strategies for enhancing fruit yield and quality in cadmium-exposed trees. *South African Journal of Botany*, *167*, 441–447. DOI: 10.1016/j.sajb.2024.02.042

Alamer, K. H., & Fayez, K. A. (2020). Impact of salicylic acid on the growth and physiological activities of parsley plants under lead toxicity. *Physiology and Molecular Biology of Plants : an International Journal of Functional Plant Biology*, *26*(7), 1361–1373. DOI: 10.1007/s12298-020-00830-1

Altaf, M. A., Naz, S., Kumar, R., Sardar, H., Nawaz, M. A., Kumar, A., Lal, P., Ahmad, R., Hayat, F., Wani, M. A., Tiwari, R. K., & Lal, M. K. (2023). Unraveling the mechanisms of cadmium toxicity in horticultural plants: Implications for plant health. *South African Journal of Botany*, *163*, 433–442. DOI: 10.1016/j.sajb.2023.10.064

Altaf, M. A., Shahid, R., Naz, S., Ahmad, R., Manzoor, M. A., Alsahli, A. A., Altaf, M. M., & Ahmad, P. (2024a). Vanadium toxicity was alleviated by supplementation of silicon in tomato seedlings: Upregulating antioxidative enzymes and glyoxalase system. *Plant Physiology and Biochemistry : PPB*, *213*, 108809. DOI: 10.1016/j.plaphy.2024.108809

Altaf, M. M., Awan, Z. A., Ashraf, S., Altaf, M. A., Zhu, Z., Alsahli, A. A., & Ahmad, P. (2024b). Melatonin induced reversibility of vanadium toxicity in muskmelon by regulating antioxidant defense and glyoxalase systems. *Journal of Hazardous Materials*, *473*, 134452. DOI: 10.1016/j.jhazmat.2024.134452

Angulo-Bejarano, P. I., Puente-Rivera, J., & Cruz-Ortega, R. (2021). Metal and metalloid toxicity in plants: An overview on molecular aspects. *Plants*, *10*(4), 635. DOI: 10.3390/plants10040635

Arteca, R. N., & Arteca, J. M. (2007). Heavy-metal-induced ethylene production in Arabidopsis thaliana. *Journal of Plant Physiology*, *164*(11), 1480–1488. DOI: 10.1016/j.jplph.2006.09.006

Asgher, M., Khan, M. I. R., Anjum, N. A., & Khan, N. A. (2015). Minimising toxicity of cadmium in plants—Role of plant growth regulators. *Protoplasma*, *252*(2), 399–413. DOI: 10.1007/s00709-014-0710-4

Ashfaque, F., Farooq, S., Chopra, P., & Chhillar, H. (2020). Improving heavy metal tolerance through plant growth regulators and osmoprotectants in plants. In *Improving Abiotic Stress Tolerance in Plants* (pp. 69–98). CRC Press. DOI: 10.1201/9780429027505-5

Asif, R., Yasmin, R., Mustafa, M., Ambreen, A., Mazhar, M., Rehman, A., Umbreen, S., & Ahmad, M. 2022. Phytohormones as plant growth regulators and safe protectors against biotic and abiotic stress. In *Plant Hormones-Recent Advances, New Perspectives and Applications*. IntechOpen. DOI: 10.5772/intechopen.102832

Asija, S., Seth, T., Umar, S., & Gupta, R. (2023). Polyamines and their crosstalk with phytohormones in the regulation of plant defense responses. *Journal of Plant Growth Regulation*, *42*(8), 5224–5246. DOI: 10.1007/s00344-022-10837-5

Askari-Khorasgani, O., & Pessarakli, M. (2019). Phytohormone homeostasis and crosstalk effects in response to osmotic stress. In *Handbook of Plant and Crop Stress* (4th ed., pp. 361–384). CRC Press.

Aslam, M. M., Waseem, M., Jakada, B. H., Okal, E. J., Lei, Z., Saqib, H. S. A., Yuan, W., Xu, W., & Zhang, Q. (2022). Mechanisms of abscisic acid-mediated drought stress responses in plants. *International Journal of Molecular Sciences*, *23*(3), 1084. DOI: 10.3390/ijms23031084

Baker, A. J., McGrath, S. P., Reeves, R. D., & Smith, J. A. C. (2020). Metal hyperaccumulator plants: a review of the ecology and physiology of a biological resource for phytoremediation of metal-polluted soils. *Phytoremediation of contaminated soil and water*, pp.85-107.

Bano, A., Singh, K., Singh, S. P., & Sharma, P. (2023). Abscisic Acid: Metabolism, signaling, and crosstalk with other phytohormones under heavy metal stress. *Stresses*, *3*(4), 665–686. DOI: 10.3390/stresses3040046

Bashmakov, D. I., Pynenkova, N. A., Sazanova, K. A., & Lukatkin, A. S. (2012). Effect of the synthetic growth regulator Cytodef and heavy metals on oxidative status in cucumber plants. *Russian Journal of Plant Physiology: a Comprehensive Russian Journal on Modern Phytophysiology*, *59*(1), 59–64. DOI: 10.1134/S1021443712010049

Basit, F., Bhat, J. A., Alyemeni, M. N., Shah, T., & Ahmad, P. (2023). Nitric oxide mitigates vanadium toxicity in soybean (Glycine max L.) by modulating reactive oxygen species (ROS) and antioxidant system. *Journal of Hazardous Materials*, *451*, 131085. DOI: 10.1016/j.jhazmat.2023.131085

Blázquez, M. A. (2024). Polyamines: Their role in plant development and stress. *Annual Review of Plant Biology*, *75*(1), 75. DOI: 10.1146/annurev-arplant-070623-110056

Bücker-Neto, L., Paiva, A. L. S., Machado, R. D., Arenhart, R. A., & Margis-Pinheiro, M. (2017). Interactions between plant hormones and heavy metals responses. *Genetics and Molecular Biology*, *40*(1, suppl 1), 373–386. DOI: 10.1590/1678-4685-gmb-2016-0087

Bulak, P., Walkiewicz, A., & Brzezińska, M. (2014). Plant growth regulators-assisted phytoextraction. *Biologia Plantarum*, *58*(1), 1–8. DOI: 10.1007/s10535-013-0382-5

Checker, V. G., Kushwaha, H. R., Kumari, P., & Yadav, S. (2018). Role of phytohormones in plant defense: signaling and cross talk. In *Molecular aspects of plant-pathogen interaction* (pp. 159–184). Springer Singapore. DOI: 10.1007/978-981-10-7371-7_7

Chen, L., Hu, W. F., Long, C., & Wang, D. (2021). Exogenous plant growth regulator alleviate the adverse effects of U and Cd stress in sunflower (Helianthus annuus L.) and improve the efficacy of U and Cd remediation. *Chemosphere*, *262*, 127809. DOI: 10.1016/j.chemosphere.2020.127809

Dar, N. A., Amin, I., Wani, W., Wani, S. A., Shikari, A. B., Wani, S. H., & Masoodi, K. Z. (2017). Abscisic acid: A key regulator of abiotic stress tolerance in plants. *Plant Gene*, *11*, 106–111. DOI: 10.1016/j.plgene.2017.07.003

Dar, S. A., Lone, R., Tyub, S., Kamili, A. N., & Nawchoo, I. A. (2023). Crosstalk of ethylene and salicylic acid in the amelioration of toxic effects of heavy metal stress in mustard. In *Plant Phenolics in Abiotic Stress Management* (pp. 173–193). Springer Nature Singapore. DOI: 10.1007/978-981-19-6426-8_9

Dubey, R. S. (2010). Metal toxicity, oxidative stress and antioxidative defense system in plants. In *Reactive oxygen species and antioxidants in higher plants* (pp. 177–203). CRC Press.

Dubey, S., Shri, M., Gupta, A., Rani, V., & Chakrabarty, D. (2018). Toxicity and detoxification of heavy metals during plant growth and metabolism. *Environmental Chemistry Letters*, *16*(4), 1169–1192. DOI: 10.1007/s10311-018-0741-8

El-Mogy, M.M., Jan, M.F., Naz, I., Ahmad, I., Ahmad, R. and Altaf, M.T., (2025). Nanoparticle Innovations for Mitigating Metal Toxicity in Plants. Phyton (0031-9457), 94(3).

Emamverdian, A., Ding, Y., Mokhberdoran, F., & Xie, Y. (2015). Heavy metal stress and some mechanisms of plant defense response. *TheScientificWorldJournal*, *2015*(1), 756120. DOI: 10.1155/2015/756120

Farman, S., Mushtaq, A., & Azeem, M. W. (2019). Plant growth regulators (PGRs) and their applications: A review. *International Journal of Chemical and Biochemical Science*, *15*, 94–103.

Fryzova, R., Pohanka, M., Martinkova, P., Cihlarova, H., Brtnicky, M., Hladky, J. and Kynicky, J., (2017). Oxidative stress and heavy metals in plants. *Reviews of environmental contamination and toxicology volume 245*, pp.129-156.

Gao, Y., Ma, X., Zhang, Z., Wang, X., & Wang, Y. (2024). Brassinolides signaling pathway: Tandem response to plant hormones and regulation under various abiotic stresses. *Horticulture Advances*, *2*(1), 27. DOI: 10.1007/s44281-024-00050-7

Gul, A. S., Naveed, F., Ali, M., Ahmad, R., & Saqib, M. (2021). Effect of different wastewater irrigation regimes on growth of mulberry (Morus macroura Miq.). *Erwerbs-Obstbau*, *63*(3), 331–337. DOI: 10.1007/s10341-021-00577-3

Hou, D., O'Connor, D., Igalavithana, A. D., Alessi, D. S., Luo, J., Tsang, D. C., Sparks, D. L., Yamauchi, Y., Rinklebe, J., & Ok, Y. S. (2020). Metal contamination and bioremediation of agricultural soils for food safety and sustainability. *Nature Reviews. Earth & Environment*, *1*(7), 366–381. DOI: 10.1038/s43017-020-0061-y

Hsu, Y. T., & Kao, C. H. (2003). Role of abscisic acid in cadmium tolerance of rice (Oryza sativa L.) seedlings. *Plant, Cell & Environment*, *26*(6), 867–874. DOI: 10.1046/j.1365-3040.2003.01018.x

Hu, B., Deng, F., Chen, G., Chen, X., Gao, W., Long, L., Xia, J., & Chen, Z. H. (2020). Evolution of abscisic acid signaling for stress responses to toxic metals and metalloids. *Frontiers in Plant Science*, *11*, 909. DOI: 10.3389/fpls.2020.00909

Huang, R., Cui, X., Luo, X., Mao, P., Zhuang, P., Li, Y., Li, Y., & Li, Z. (2021). Effects of plant growth regulator and chelating agent on the phytoextraction of heavy metals by Pfaffia glomerata and on the soil microbial community. *Environmental Pollution*, *283*, 117159. DOI: 10.1016/j.envpol.2021.117159

Ilieva, I., & Sainova, I. (2023). Mechanisms of Action of Heavy Metals, Related with Abnormal Protein and Enzyme Activity in Male Infertility Aspect. *Acta Morphologica et Anthropologica, 30*(1-2), 1–2. DOI: 10.7546/AMA.30.1-2.2023.09

Jan, S., Bhardwaj, R., Sharma, N. R., & Singh, R. (2024). Unraveling the role of plant growth regulators and plant growth promoting rhizobacteria in phytoremediation. *Journal of Plant Growth Regulation, 43*(8), 2471–2487. DOI: 10.1007/s00344-024-11284-0

Johnson, A., Singhal, N., & Hashmatt, M. (2011). Metal–plant interactions: toxicity and tolerance. In *Biomanagement of metal-contaminated soils* (pp. 29–63). Springer. DOI: 10.1007/978-94-007-1914-9_2

Kapoor, D., Bhardwaj, S., Gautam, S., Rattan, A., Bhardwaj, R., & Sharma, A. (2022). Brassinosteroids in plant nutrition and heavy metal tolerance. In *Brassinosteroids in plant developmental biology and stress tolerance* (pp. 217–235). Academic Press. DOI: 10.1016/B978-0-12-813227-2.00008-4

Keramat, B., Kalantari, K. M., & Arvin, M. J. (2010). Effects of methyl jasmonate treatment on alleviation of cadmium damages in soybean. *Journal of Plant Nutrition, 33*(7), 1016–1025. DOI: 10.1080/01904161003728685

Keunen, E., Schellingen, K., Vangronsveld, J., & Cuypers, A. (2016). Ethylene and metal stress: Small molecule, big impact. *Frontiers in Plant Science, 7*, 23. DOI: 10.3389/fpls.2016.00023

Khan, M. I. R., Syeed, S., Nazar, R., & Anjum, N. A. (2012). An insight into the role of salicylic acid and jasmonic acid in salt stress tolerance. *Phytohormones and abiotic stress tolerance in plants*, pp.277-300.

Khan, N., Bano, A. M., & Babar, A. (2020). Impacts of plant growth promoters and plant growth regulators on rainfed agriculture. *PLoS One, 15*(4), e0231426. DOI: 10.1371/journal.pone.0231426

Kour, J., Kohli, S. K., Khanna, K., Bakshi, P., Sharma, P., Singh, A. D., Ibrahim, M., Devi, K., Sharma, N., Ohri, P., Skalicky, M., Brestic, M., Bhardwaj, R., Landi, M., & Sharma, A. (2021). Brassinosteroid signaling, crosstalk and, physiological functions in plants under heavy metal stress. *Frontiers in Plant Science, 12*, 608061. DOI: 10.3389/fpls.2021.608061

Kumar, S., Shah, S. H., Vimala, Y., Jatav, H. S., Ahmad, P., Chen, Y., & Siddique, K. H. (2022). Abscisic acid: Metabolism, transport, crosstalk with other plant growth regulators, and its role in heavy metal stress mitigation. *Frontiers in Plant Science*, *13*, 972856. DOI: 10.3389/fpls.2022.972856

Lan, X., Li, J., Chen, J., Liu, J., Cao, F., Liao, C., Zhang, Z., Gu, M., Wei, Y., Shen, F., Wei, X., Luo, X., & Zhang, X. (2023). Effects of foliar applications of Brassinolide and Selenium on the accumulation of Arsenic and Cadmium in rice grains and an assessment of their health risk. *International Journal of Phytoremediation*, *25*(2), 161–171. DOI: 10.1080/15226514.2022.2066064

Maleki, M., Ghorbanpour, M., & Kariman, K. (2017). Physiological and antioxidative responses of medicinal plants exposed to heavy metals stress. *Plant Gene*, *11*, 247–254. DOI: 10.1016/j.plgene.2017.04.006

Mansoor, S., Ali, A., Kour, N., Bornhorst, J., AlHarbi, K., Rinklebe, J., Abd El Moneim, D., Ahmad, P., & Chung, Y. S. (2023). Heavy metal induced oxidative stress mitigation and ROS scavenging in plants. *Plants*, *12*(16), 3003. DOI: 10.3390/plants12163003

Manzoor, H., Mehwish, , Bukhat, S., Rasul, S., Rehmani, M. I. A., Noreen, S., Athar, H.-R., Zafar, Z. U., Skalicky, M., Soufan, W., Brestic, M., Habib-ur-Rahman, M., Ogbaga, C. C., & EL Sabagh, A. (2022). Methyl jasmonate alleviated the adverse effects of cadmium stress in pea (Pisum sativum L.): A nexus of photosystem II activity and dynamics of redox balance. *Frontiers in Plant Science*, *13*, 860664. DOI: 10.3389/fpls.2022.860664

Manzoor, M., Korolev, K. and Ahmad, R., (2025). Zinc oxide nanoparticles: abiotic stress tolerance in fruit crops focusing on sustainable production. Phyton (0031-9457), 94(5).

Massoud, B. M., Sakouhi, L., & Chaoui, A. (2019). Effect of plant growth regulators, calcium and citric acid on copper toxicity in pea seedlings. *Journal of Plant Nutrition*, *42*(10), 1230–1242. DOI: 10.1080/01904167.2019.1609506

Mondal, S., Pramanik, K., Ghosh, S. K., Pal, P., Ghosh, P. K., Ghosh, A., & Maiti, T. K. (2022). Molecular insight into arsenic uptake, transport, phytotoxicity, and defense responses in plants: A critical review. *Planta*, *255*(4), 87. DOI: 10.1007/s00425-022-03869-4

Mousavi, S. R., Niknejad, Y., Fallah, H., & Tari, D. B. (2020). Methyl jasmonate alleviates arsenic toxicity in rice. *Plant Cell Reports*, *39*(8), 1041–1060. DOI: 10.1007/s00299-020-02547-7

Muhammad, H. M. D., Naz, S., Lal, M. K., Tiwari, R. K., Ahmad, R., Nawaz, M. A., Das, R., & Altaf, M. A. (2024). Melatonin in business with abiotic stresses in vegetable crops. *Scientia Horticulturae*, *324*, 112594. DOI: 10.1016/j.scienta.2023.112594

Naz, S., Anjum, M. A., Sadiq, B., Ahmad, R., Altaf, M. A., El-Sheikh, M. A., & Shakoor, A. (2022). Purification of sewage wastewater though sand column filter for lessening of heavy metals accumulation in lettuce, carrot, and cauliflower. *Water (Basel)*, *14*(22), 3770. DOI: 10.3390/w14223770

Nguyen, T. Q., Sesin, V., Kisiala, A., & Emery, R. N. (2021). Phytohormonal roles in plant responses to heavy metal stress: Implications for using macrophytes in phytoremediation of aquatic ecosystems. *Environmental Toxicology and Chemistry*, *40*(1), 7–22. DOI: 10.1002/etc.4909

Nowicka, B. (2022). Heavy metal–induced stress in eukaryotic algae—Mechanisms of heavy metal toxicity and tolerance with particular emphasis on oxidative stress in exposed cells and the role of antioxidant response. *Environmental Science and Pollution Research International*, *29*(12), 16860–16911. DOI: 10.1007/s11356-021-18419-w

Nwankwo, C. E. I., Okeke, E. S., Umeoguaju, F. U., Ejeromedoghene, O., Adedipe, D. T., & Ezeorba, T. P. C. (2025). Addressing emerging contaminants in agriculture affecting plant–soil interaction: A review on bio-based and nano-enhanced strategies for soil health and global food security (GFS). *Discover Toxicology*, *2*(1), 1–30. DOI: 10.1007/s44339-025-00018-w

Ogugua, U. V., Kanu, S. A., & Ntushelo, K. (2022). Gibberellic acid improves growth and reduces heavy metal accumulation: A case study in tomato (*Solanum lycopersicum* L.) seedlings exposed to acid mine water. *Heliyon*, *8*(12), e12399. DOI: 10.1016/j.heliyon.2022.e12399

Oladimeji, T. E., Oyedemi, M., Emetere, M. E., Agboola, O., Adeoye, J. B., & Odunlami, O. A. (2024). Review on the impact of heavy metals from industrial wastewater effluent and removal technologies. *Heliyon*, *10*(23), e40370. DOI: 10.1016/j.heliyon.2024.e40370

Pál, M., Janda, T., & Szalai, G. (2018). Interactions between plant hormones and thiol-related heavy metal chelators. *Plant Growth Regulation*, *85*(2), 173–185. DOI: 10.1007/s10725-018-0391-7

Pan, W., Lu, Q., Xu, Q. R., Zhang, R. R., Li, H. Y., Yang, Y. H., Liu, H. J., & Du, S. T. (2019). Abscisic acid-generating bacteria can reduce Cd concentration in pakchoi grown in Cd-contaminated soil. *Ecotoxicology and Environmental Safety*, *177*, 100–107. DOI: 10.1016/j.ecoenv.2019.04.010

Pardo-Hernández, M., López-Delacalle, M., Martí-Guillen, J. M., Martínez-Lorente, S. E., & Rivero, R. M. (2021). ROS and NO phytomelatonin-induced signaling mechanisms under metal toxicity in plants: A review. *Antioxidants*, *10*(5), 775. DOI: 10.3390/antiox10050775

Peñarrubia, L., Romero, P., Carrió-Seguí, A., Andrés-Bordería, A., Moreno, J., & Sanz, A. (2015). Temporal aspects of copper homeostasis and its crosstalk with hormones. *Frontiers in Plant Science*, *6*, 255. DOI: 10.3389/fpls.2015.00255

Per, T. S., Khan, N. A., Masood, A., & Fatma, M. (2016). Methyl jasmonate alleviates cadmium-induced photosynthetic damages through increased S-assimilation and glutathione production in mustard. *Frontiers in Plant Science*, *7*, 1933. DOI: 10.3389/fpls.2016.01933

Piotrowska-Niczyporuk, A., Bajguz, A., Zambrzycka-Szelewa, E., & Bralska, M. (2018). Exogenously applied auxins and cytokinins ameliorate lead toxicity by inducing antioxidant defence system in green alga Acutodesmus obliquus. *Plant Physiology and Biochemistry : PPB*, *132*, 535–546. DOI: 10.1016/j.plaphy.2018.09.038

Prasad, R. (2022). Cytokinin and its key role to enrich the plant nutrients and growth under adverse conditions-an update. *Frontiers in Genetics*, *13*, 883924. DOI: 10.3389/fgene.2022.883924

Rahman, S. U., Li, Y., Hussain, S., Hussain, B., Khan, W. U. D., Riaz, L., Ashraf, M. N., Khaliq, M. A., Du, Z., & Cheng, H. (2023). Role of phytohormones in heavy metal tolerance in plants: A review. *Ecological Indicators*, *146*, 109844. DOI: 10.1016/j.ecolind.2022.109844

Rajewska, I., Talarek, M., & Bajguz, A. (2016). Brassinosteroids and response of plants to heavy metals action. *Frontiers in Plant Science*, *7*, 629. DOI: 10.3389/fpls.2016.00629

Rathod, S. V., Saras, P., & Gondaliya, S. M. (2024). Environmental pollution: Threats and challenges for management. In *Eco-restoration of polluted environment* (pp. 1–34). CRC Press. DOI: 10.1201/9781003423393-1

Rostami, S., & Azhdarpoor, A. (2019). The application of plant growth regulators to improve phytoremediation of contaminated soils: A review. *Chemosphere*, *220*, 818–827. DOI: 10.1016/j.chemosphere.2018.12.203

Roychowdhury, R., Khan, M. H., & Choudhury, S. (2019). Physiological and molecular responses for metalloid stress in rice—A Comprehensive Overview. *Advances in rice research for abiotic stress tolerance*, pp.341-369.

Rucińiska-Sobkowiak, R. (2010). Oxidative stress in plants exposed to heavy metals. *Postępy Biochemii*, *56*(2), 191–200.

Sharma, A., Sidhu, G. P. S., Araniti, F., Bali, A. S., Shahzad, B., Tripathi, D. K., Brestic, M., Skalicky, M., & Landi, M. (2020). The role of salicylic acid in plants exposed to heavy metals. *Molecules (Basel, Switzerland)*, *25*(3), 540. DOI: 10.3390/molecules25030540

Singh, A. D., Sharma, P., Kohli, S. K., Kumar, P., Singh, R., Arora, P., Sharma, P., Kaur, R., Sharma, A., & Bhardwaj, R. (2020). Role of plant growth regulators (PGRs) in mitigation of heavy metal phytotoxicity in plants. In *Cellular and molecular phytotoxicity of heavy metals* (pp. 263–304). Springer International Publishing. DOI: 10.1007/978-3-030-45975-8_14

Singh, I., & Shah, K. (2014). Exogenous application of methyl jasmonate lowers the effect of cadmium-induced oxidative injury in rice seedlings. *Phytochemistry*, *108*, 57–66. DOI: 10.1016/j.phytochem.2014.09.007

Singhal, R. K., Kumar, M., Bose, B., Mondal, S., Srivastava, S., Dhankher, O. P., & Tripathi, R. D. (2023). Heavy metal (loid) s phytotoxicity in crops and its mitigation through seed priming technology. *International Journal of Phytoremediation*, *25*(2), 187–206. DOI: 10.1080/15226514.2022.2068502

Sirhindi, G., Mir, M. A., Abd-Allah, E. F., Ahmad, P., & Gucel, S. (2016). Jasmonic acid modulates the physio-biochemical attributes, antioxidant enzyme activity, and gene expression in Glycine max under nickel toxicity. *Frontiers in Plant Science*, *7*, 591. DOI: 10.3389/fpls.2016.00591

Song, C., Bibi, Z., Ali, S., & Ahmad, R. (2026). Silicon mitigates aluminum toxicity in lettuce by improving photosynthetic pigments and antioxidant defense system. *Biometals : An International Journal on the Role of Metal Ions in Biology, Biochemistry, and Medicine*, •••, 1–14. DOI: 10.1007/s10534-026-00819-5

Steffens, B. (2014). The role of ethylene and ROS in salinity, heavy metal, and flooding responses in rice. *Frontiers in Plant Science*, *5*, 685. DOI: 10.3389/fpls.2014.00685

Sytar, O., Kumari, P., Yadav, S., Brestic, M., & Rastogi, A. (2019). Phytohormone priming: Regulator for heavy metal stress in plants. *Journal of Plant Growth Regulation*, *38*(2), 739–752. DOI: 10.1007/s00344-018-9886-8

Tang, Y., Wang, L., Xie, Y., Yu, X., Li, H., Lin, L., Liao, M. A., Wang, Z., Sun, G., Wang, X., Liang, D., Xia, H., & Tu, L. (2020). Effects of exogenous abscisic acid on the growth and cadmium accumulation of lettuce under cadmium-stress conditions. *International Journal of Environmental Analytical Chemistry*, *100*(6), 720–731. DOI: 10.1080/03067319.2019.1639686

Tassi, E., Pouget, J., Petruzzelli, G., & Barbafieri, M. (2008). The effects of exogenous plant growth regulators in the phytoextraction of heavy metals. *Chemosphere*, *71*(1), 66–73. DOI: 10.1016/j.chemosphere.2007.10.027

Thao, N. P., Khan, M. I. R., Thu, N. B. A., Hoang, X. L. T., Asgher, M., Khan, N. A., & Tran, L. S. P. (2015). Role of ethylene and its cross talk with other signaling molecules in plant responses to heavy metal stress. *Plant Physiology*, *169*(1), 73–84. DOI: 10.1104/pp.15.00663

Ullah, A., Hazrat, A., Khan, B. A., Saqib, S., & Ullah, F. (2025). Mitigating lead stress in barley using gibberellic acid (GA3): Effects on morpho-physiological and biochemical parameters. *Journal of Plant Growth Regulation*, *44*(7), 1–12. DOI: 10.1007/s00344-025-11654-2

Wang, Y., Mostafa, S., Zeng, W., & Jin, B. (2021). Function and mechanism of jasmonic acid in plant responses to abiotic and biotic stresses. *International Journal of Molecular Sciences*, *22*(16), 8568. DOI: 10.3390/ijms22168568

Wani, A. B., Chadar, H., Wani, A. H., Singh, S., & Upadhyay, N. (2017). Salicylic acid to decrease plant stress. *Environmental Chemistry Letters*, *15*(1), 101–123. DOI: 10.1007/s10311-016-0584-0

Witkowska, D., Słowik, J., & Chilicka, K. (2021). Heavy metals and human health: Possible exposure pathways and the competition for protein binding sites. *Molecules (Basel, Switzerland)*, *26*(19), 6060. DOI: 10.3390/molecules26196060

Yan, Z., Zhang, W., Chen, J., & Li, X. (2015). Methyl jasmonate alleviates cadmium toxicity in *Solanum nigrum* by regulating metal uptake and antioxidative capacity. *Biologia Plantarum*, *59*(2), 373–381. DOI: 10.1007/s10535-015-0491-4

Yang, W., Dai, H., Wei, S., & Skuza, L. (2024). The effect of exogenous plant growth regulators on elevated Cd phytoremediation by Solanum nigrum L. in contaminated soil. *Environmental Science and Pollution Research International*, *31*(3), 3964–3975. DOI: 10.1007/s11356-023-31420-9

Yu, X., Yang, L., Fan, C., Hu, J., Zheng, Y., Wang, Z., Liu, Y., Xiao, X., Yang, L., Lei, T., Jiang, M., Jiang, B., Pan, Y., Li, X., Gao, S., & Zhou, Y. (2023). Abscisic acid (ABA) alleviates cadmium toxicity by enhancing the adsorption of cadmium to root cell walls and inducing antioxidant defense system of Cosmos bipinnatus. *Ecotoxicology and Environmental Safety*, *261*, 115101. DOI: 10.1016/j.ecoenv.2023.115101

Zaid, A., Ahmad, B., Jaleel, H., Wani, S. H., & Hasanuzzaman, M. (2020). A critical review on iron toxicity and tolerance in plants: role of exogenous phytoprotectants. *Plant micronutrients: Deficiency and toxicity management*, pp.83-99.

Zhang, L., Liu, Z., Song, Y., Sui, J., & Hua, X. (2024). Advances in the involvement of metals and metalloids in plant defense response to external stress. *Plants*, *13*(2), 313. DOI: 10.3390/plants13020313

Zhao, Y., Wang, J., Huang, W., Zhang, D., Wu, J., Li, B., Li, M., Liu, L., & Yan, M. (2023). Abscisic-acid-regulated responses to alleviate cadmium toxicity in plants. *Plants*, *12*(5), 1023. DOI: 10.3390/plants12051023

Zulfiqar, F., & Ashraf, M. (2021). Bioregulators: Unlocking their potential role in regulation of the plant oxidative defense system. *Plant Molecular Biology*, *105*(1), 11–41. DOI: 10.1007/s11103-020-01077-w

# Chapter 3
# Biostimulants' Role in Alleviation of Metal-Induced Toxicity in Plants

**Hafiza Muniba Din Muhammad**
http://orcid.org/0000-0003-1771-3624
*Bahauddin Zakariya University, Multan, Pakistan*

## ABSTRACT

*Climate change is considered as one of the major constrain reducing the productivity of crops. The soil contamination is increasing gradually due to the several factors such as brackish water, urbanization, and industrial effluents. Phytoremediation is considered as effective remedy for the removal toxic substances form the soil focusing on sustainable growth and higher yield. However, current chapter aimed to evaluate the impact of biostimulants in mitigation of metal toxicity in plants. Biostimulants are widely used in agriculture to increase crop yield and resilience against various stresses. Moreover, recent research has shown that biostimulants can lessen the harmful effects of contaminants on plants, boosting some species' phytoremediation effectiveness. Biostimulants composed of humus, protein and amino acid hydrolysate, inorganic salts, microbes, sea-*

DOI: 10.4018/979-8-3693-9899-9.ch003

*weed, plant extracts, and fungi are used on crops against metal toxicity. Conclusively, biostimulants are effective way to cope adverse effects of metal toxicity in plants.*

## INTRODUCTION

Heavy metals are continuously released into the environment by human activities such as burning fossil fuels, mining and smelting metal ores, growing cities and industries, producing a lot of municipal trash, and farming practices (Srikanth et al., 2022). Heavy metals are thought to be the primary contaminants in groundwater. The toxic metals in plants have an impact on plant growth and crop yield and are considered a serious problem for the environment as well as agriculture (Sardar et al., 2013). Metals can cause oxidative harm, stunt plant growth, disrupt photosynthetic activity, substitute other metals for pigments, and hasten senescence in plants (Alengebawy et al., 2021). Metal excess can cause the production of reactive oxygen species (ROS), which can result in oxidative injury and cell destruction (Altaf et al., 2024). Plants use several kinds of strategies to combat heavy metal toxicity, such as antioxidant machinery to combat ROS (Bayata, 2020). Some proteins are essential for controlling the way plants react to metal toxicity. The basic composition of the soil, human activity such as mining, and reducing the pH of the soil can all lead to toxic metals (Ullah et al., 2022). Developing methods to reduce toxic metals in plants and enhance crop productivity in metal-contaminated areas requires a comprehension of these routes (DalCorso et al., 2013).

The pesticides have been found in soil and water at levels that are frequently beyond the permissible limits (Altaf et al., 2023). To keep pesticides from getting into the food chain and posing serious risks to the health of people and animals, it is important to minimize their buildup in soil and water (Kaur et al., 2024). Although pesticides are used in agriculture to reduce and manage weed competition among crops, their ability to selectively target weeds is sometimes just theoretical. Several pesticides have the ability to infect non-target organisms, such as crops, causing changes in their morphology, physiology, and biology (Tahir et al., 2024). For instance, pesticides can lower crop growth and biomass production, disrupt mineral nutrition and utilization efficiency, lower the concentration of chlorophyll, and induce cell death (Kalyabina et al., 2021; Raffa and Chiampo, 2021). Hence, pesticides are involved in soil contamination.

The elimination of emissions of pollutants into the environment is in need of time (Medvedkina & Khodochenko, 2020). Another crucial component in lowering environmental pollution is the transition to renewable energy sources (Ahmad et al., 2024). To evaluate their impact on air, soil, and water ecosystems and to decide what steps and actions should be taken, it is crucial to model both point and non-point sources of pollution (Loague and Corwin, 2006). Particular focus needs to be placed on cleaning up contaminated areas in addition to critical emissions. The remediation techniques comprise dig and dump, biochemical oxidation process, incineration, thermal treatment, dilution of the contaminants, electrokinetic, soil flushing, excavation, and disposal that can presently be used to clean hazardous environments (Wang et al., 2022). However, these techniques can be costly, energy-intensive, require specialized equipment, and have the potential to harm the environment and result in secondary pollution (Sharma et al., 2025). For instance, the majority of the physical or chemical methods listed above have an impact on the biological activity, structure, and fertility of polluted soil. Moreover, the contamination may not be eliminated (Sah et al., 2025).

Plant biostimulants, also known as biofertilizers, biostimulators, plant probiotics, and metabolic enhancers, widely used in agriculture, have the potential to increase plant quality and productivity (Carillo et al., 2025). Their use in crops actually enables the induction of physiological and molecular processes that can enhance crop production and product quality (Bint Tariq and Umer, 2025). Fertilizing products whose function is to stimulate plant nutrition processes independently of the product's nutrient content to improve the following aspects of the plant and plant rhizosphere, such as nutrient use efficiency, tolerance resistance to metal stress, and accessibility of the restricted nutrients in the soil, are due to the application of biostimulants (Khoulati et al., 2025).

Biostimulants do not belong to the category of fertilizers and plant protection products; their main purpose is neither to supply nutrients and shield plants from pests and diseases (Buss et al., 2025). These biostimulants can improve the availability of nutrients to plant roots, increase the plant's potential to use nutrients, and cope with metal toxicity (Lanrewaju and Fadiji, 2025). Additionally, Biostimulants effectively enhance plant nutrient uptake, which may further allow the decrease in the use of chemical fertilizers that are frequently used in agriculture (Suman et al., 2025). Due to the potential decrease in the quantity of synthetic substances that this activity consumes, this could also support the environmental sustainability of agriculture. Biostimulants have been employed to promote positive effects in several crops (Alfonso et al., 2025). Therefore, the current chapter aims to determine the positive role of biostimulants in the enhancement of crop growth and yield growing under polluted soils.

## UNDERSTAND THE ADVERSE EFFECTS OF METAL TOXICITY ON PLANT PRODUCTIVITY

Plants grown on soils contaminated with heavy metals experience a decrease in growth due to modifications in physiological and biochemical processes (Yadav et al., 2025). The disruption of nutrient uptake and translocation, photosynthesis impairment, photosynthetic pigments reduction, inhibiting electron transport, decreasing $CO_2$ fixation, disorganizing chloroplasts, and causing photooxidative damage, the production of ROS, the inhibition of antioxidative enzymes, cellular redox imbalance, DNA damage, and protein oxidation are some of the phytotoxic indices of heavy metals (Muhammad et al., 2024; Derrick et al., 2025). It will be possible to evaluate heavy metal-tolerant plant species and their application for the phytoremediation of contaminated soils by evaluating the effects of heavy metal toxicity on plants (Shivappa et al., 2025). The stress produced from excess metals is a crucial component that adversely affects agricultural production (Figure 1). It is commonly recognized that metals are a component of soil. However, the ecology may suffer if they are present in large enough quantities in plants and soil (Hasan and Tarannum, 2025). Heavy metals can go up the food chain, eventually building up in the bodies of living organisms and potentially endangering their health after being absorbed by plants (Shaikh et al., 2025). When heavy metals enter into groundwater, they can either be absorbed by plants or released into the environment as atmospheric gases. Alternatively, they might develop semi-permanent relationships with organic materials in the soil, which could eventually affect human health (Ochoa et al., 2025). To maintain sustainability, soil contamination levels must be kept below the recommended toxicity thresholds (Manna et al., 2025). Chemical fertilizers and pesticides have been shown to increase the risk of metal contamination of the soil, which can then build up in the tissue of crops grown in contaminated soil. The amount of total organic matter and pH levels in soil has a major impact on the mobility of metals. Metals are introduced to soil, and their ability to migrate downward is restricted by either the soil's ability to hold the metals or the metals' interaction with the nearby waste matrix, which may increase their mobility (Shar et al., 2025). Legumes' ability to develop, thrive, and fix nitrogen may be negatively impacted by high soil metal concentrations (Hasan et al., 2025). The biological process by which microorganisms absorb or convert heavy metals into less hazardous compounds is known as microbial remediation (Naz et al., 2022). To treat heavy metal contamination in soil, microbial remediation is a method that mainly uses various mechanisms, including biosorption, bioaccumulation, bioleaching, biovolatilization, and biomineralization (Muhammad et al., 2025). Metal toxicity showed the adverse effects on plants, focusing on growth and yield characteristics (Table 1).

*Figure 1. Adverse effects of metal toxicity in plants in terms of growth and productivity*

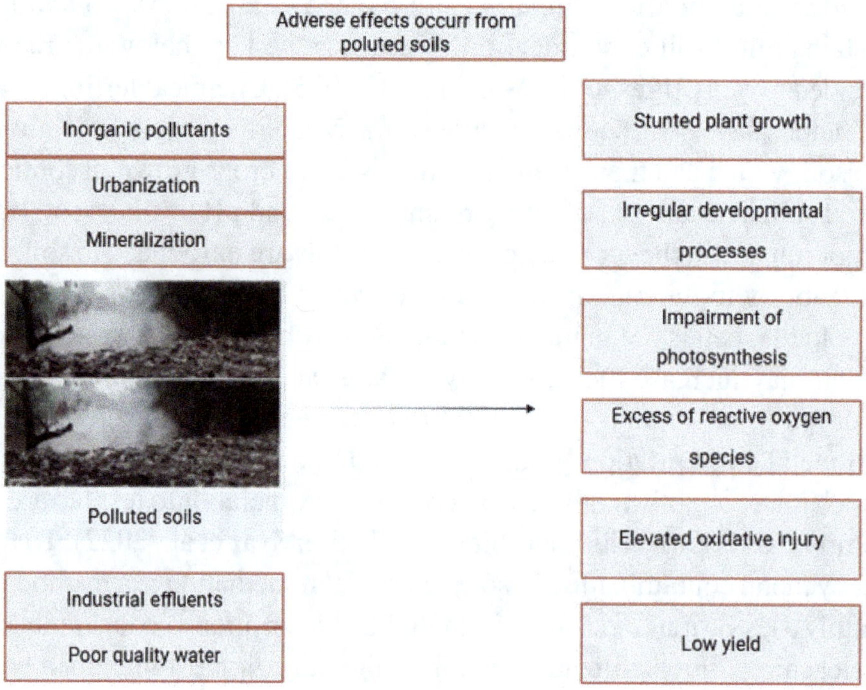

*Figure 2. Different biostimulants commonly used in plants against metal toxicity*

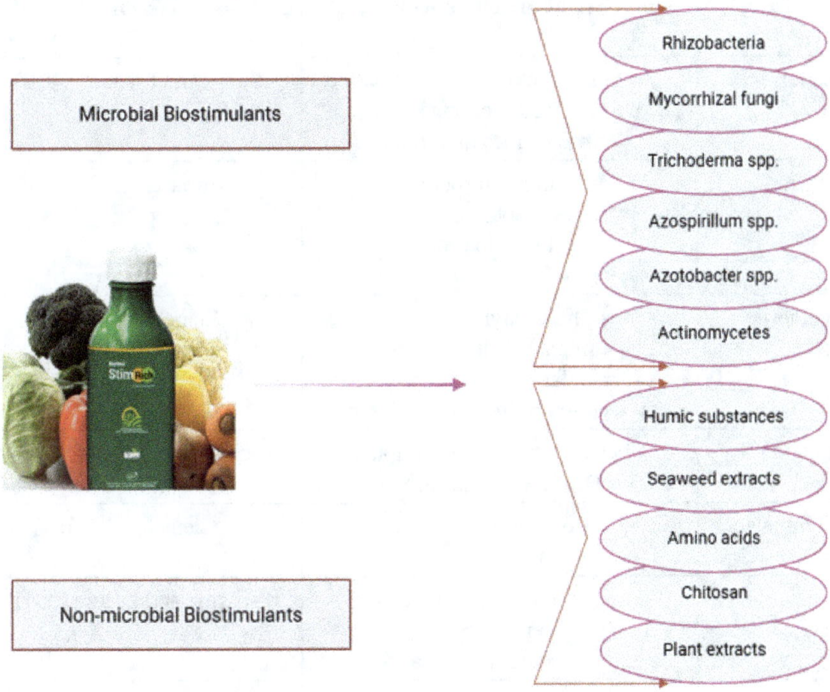

*Table 1. Metal toxicity illustrates the adverse effects on plant growth and production*

| Characteristic affected | Metal toxicity effects | References |
|---|---|---|
| Germination | ▪ Inhibition of seed germination<br>▪ Reduced seedlings growth<br>▪ Enzymes disruption | Subin and Francis (2013) |
| Root growth | ▪ Diminished root growth<br>▪ Root hair loss<br>▪ Injury to root tips<br>▪ Poor anchorage | Anwar et al. (2021) |
| Photosynthesis process | ▪ Chlorophyll degradation<br>▪ Impaired photosynthesis<br>▪ Irregular stomatal conductance<br>▪ Disruption in transpiration rate | Al-Charchafchi and Al-Quadan (2006) |
| Water relations | ▪ Disruption of water uptake<br>▪ Distress transpiration rate | Rucińska-Sobkowiak (2016) |
| Nutrient imbalance | ▪ Restriction of nutrients assimilation | Clemens (2006) |
| Enzymatic activities | ▪ Inactivation of metabolic enzymes<br>▪ Metabolic imbalances | Sharma and Dietz (2009) |
| Hormonal disruptions | ▪ Disruption of phytohormones<br>▪ Stunted growth | Biswas and Pal (2025) |
| Yield constraints | ▪ Interruption in flower and fruit formation<br>▪ Reduction in seed quality | Emamverdian et al. (2020) |
| Oxidative stress | ▪ Cell membrane impairment<br>▪ Lipid peroxidation | Choudhary and Kumar (2025) |

## DIFFERENT BIOSTIMULANTS TO ALLEVIATE TOLERANCE MECHANISMS IN PLANTS SUBJECTED TO METAL TOXICITY

The cellular ROS balance is crucial for plants because ROS has two functions. Once their concentration rises under stress, they have a negative impact on cellular metabolism while serving as signaling molecules that control plant growth and development and help plants adapt to stress (Rao et al., 2025). As a result, plants have an antioxidant system that controls the balance of cellular redox potential. However, antioxidative protection's influence and capacity may vary greatly depending upon the species and genotype (Srivastava et al., 2025). Enzymatic and non-enzymatic components make up plant antioxidative defense, which scavenges ROS oxidative action to stop cellular harm (Choudhary and Kumar, 2025). Superoxide dismutase (SOD), catalase (CAT), ascorbate peroxidase (APX), non-specific peroxidase (POX), monodehydroascorbate reductase (MDHAR), dehydroascorbate reductase (DHAR), glutathione S-transferase (GST), glutathione peroxidase (GPX), alternative oxidase (AOX), and peroxiredoxin (PRX) make up the enzymatic antioxidant defense, whereas the non-enzymatic antioxidant defense system is made up of ascorbic acid (AsA), glutathione (GSH), carotenoids, flavonoids, and α-tocopherol (Basit et al., 2025). A key component of defense against ROS is enzymatic defense. SOD's activity converts superoxide radicals into $H_2O_2$, whereas APX, GPX, and CAT are in control of further converting them into water. APX exhibits its activity in the cytosol, chloroplasts, mitochondria, and peroxisomes through the ascorbate-glutathione (AsA-GSH) cycle (Dhiman et al., 2025). Apoplasts are among the nearly every cellular compartment that contains AsA, a potent non-enzymatic antioxidant. GSH is significant because it contributes to AsA regeneration through the AsA-GSH cycle (Sreelatha et al., 2025). The thylakoid membrane and other cellular membranes are protected against lipid peroxidation by carotenoids, phenols, and α-tocopherol (Manzoor et al., 2025; Hasan and Naushin, 2026). Therefore, crop health under metal toxicity depends on a slight balance between ROS production and antioxidant activity (El-Mogy et al., 2025; Biswas and Pal, 2025).

Seaweed, chitosan, plant-derived bioactive compounds, polyphenols, and allelochemicals are many types of extracts that come from various species. Seaweeds have long been utilized as a source of organic matter and fertilizers, but more recently, their efficacy as biostimulants has been approved (Thakur et al., 2025). Seaweed extracts have a variety of application techniques, such as foliar treatments and root zone application, and are rich in active minerals and organic compounds that effectively support plant growth, photosynthetic activity, and metal stress tolerance (Arikan-Algul et al., 2025). Hence, they support heavy metal fixation, soil remediation, gel formation, water retention, and soil aeration. Seaweeds regulate antioxidant activity and endogenous stress-responsive gene expression in plants, resulting in the amelioration of metal tolerance (Sharma et al., 2025). For instance, nitrate reductase activity was elevated in creeping bent grass and brinjal when seaweeds were applied as a foliar spray, stimulating nitrogen assimilation (Huang et al., 2025). Thus, seaweed extract may improve plant nutrition and help in the growth and development of plants. These extracts promote a rapid plant recovery from metal stress and enhance oxidative stress tolerance by increasing membrane stability and ROS scavenging through their cofactor function in antioxidant activity (Adnan et al., 2025). Moreover, improvements in total soluble sugar levels under metal stress have demonstrated the function of seaweeds as osmoprotectants (Čmiková et al., 2022).

The increased phenolic content and flavonoids have also been shown to contribute to seaweeds-based metal tolerance (Khan et al., 2015). Applying 100 ppm of chitosan topically to soybeans may be a useful strategy for enhancing their resistance to nickel toxicity. Therefore, the most successful treatment was foliar spray at 100 ppm (Einhardt et al., 2021; Oliveira et al., 2024). Seaweeds improved the growth of wheat seedlings and lessened the negative effects of cadmium (Uruç Parlak, 2016). Chitosan dramatically lowered the levels of cadmium in wheat roots and shoots. Under cadmium stress, chitosan increased the activity of antioxidant enzymes in wheat leaves. Plants under cadmium stress had their osmotic adjustment matter controlled by exogenous chitosan (Rasheed et al., 2020; Liu et al., 2021). The contribution of various biostimulants in enhancing tolerance against metal toxicity in plants (Table 2).

*Table 2. The contribution of various biostimulants in enhancing tolerance against metal toxicity in plants*

| Biostimulants Type | Apparatus of action | References |
|---|---|---|
| Humic constituents | Chelation of metals, enhanced nutrient uptake, and antioxidant enzyme encouragement | Ondrasek et al. (2018) |
| Seaweeds | Boosts antioxidant system, moderates stress-linked hormones, and purifies toxic substances | El Khattabi et al. (2023) |
| Protein hydrolysates | Enhances nitrogen assimilation and the antioxidant defense system | Pasković et al. (2024) |
| Mycorrhizal fungi | Metal sequestration, enhanced phosphorus uptake, and compacted metal translocation | Dhalaria et al. (2020) |
| Chitosan | Prompts defense reaction, higher phenolics and flavonoids, and inhibits metal access | Rasheed et al. (2020) |
| Biochar | Immobilizes metals, maintains soil pH, and boosts microbial action | Haiying et al. (2022) |

## HUMIC SUBSTANCES' INVOLVEMENT IN THE REDUCTION OF METAL TOXICITY IN PLANTS

Biostimulants include humic compounds, amino acids, fatty acids, and organic acids. The majority of humic compounds are naturally occurring soil organic matter that is formed by microbial activity as well as the breakdown process (Schutzendubel and Polle, 2002). Humic compounds promote plant growth and stress tolerance by improving water and nutrient uptake (Haghighi et al., 2010). The effects of humic compounds on plant physiology are not well understood, but the medium's characteristics have a significant impact on their bioactivity (Sharma and Dietz, 2009). These compounds effectively increase plants' resistance to metal stress by enhancing endogenous cytokinin, antioxidant capability, and hydration status (Ondrasek et al., 2018). Higher accumulations of minerals are linked to improved metal tolerance, while little accumulations of harmful metals are caused by humic compounds (Ghuge et al., 2023). Humic compounds are also responsible for the adjustment of ion homeostasis and the rise in proline content with a decrease in membrane leakage (Özkay et al., 2016).

A combination of amino acids, peptides, polypeptides, and denatured proteins, protein hydrolysate is formed by enzymatic, chemical, and thermal hydrolysis in plants (Perelomov et al., 2018). These preparations can be used for root and leaf fertilization and come in either powdered or granular form as well as liquid extracts. Protein hydrolysate products can increase a plant's resistance to environmental stressors. Moreover, protein hydrolysate application facilitates the easy use of amino acids and peptides for carbon and nitrogen by increasing soil respiration, biomass buildup, and microbial activity (Rong et al., 2020). Both increased plant nutrient assimilation and protein hydrolysate supplementation upregulate enzymatic activity related to carbon metabolism and nitrate assimilation (Dalvi and Bhalerao, 2013). Their positive effects on improving water status, proline content, phenolic compound accumulation, stomatal conductance, and photosynthetic efficiency have been extensively described. Increased tolerance is also a result of stress-responsive genes (Song et al., 2026). Furthermore, protein hydrolysate improves metal chelation, ROS scavenging capabilities, and antioxidant activity (Carrasco-Castilla et al., 2012). Through improved microbial activity, metal chelation, improved plant nutrient uptake, and metabolic alterations like phenolic compounds, proline synthesis, and nitrate metabolism linked to plant stress tolerance, acids directly improve soil structure (Chai et al., 2015).

## BIOSTIMULANTS AND PHYTOHORMONAL INTERPLAY IN PLANTS EXPOSED TO METAL-INDUCED CONDITIONS

The most effective and promising method for addressing yield losses brought on by different stresses—which are made worse by climate change—is the use of natural biostimulants. Numerous different chemicals included in biostimulants help plants develop more quickly, reduce stress-induced growth inhibitions, and boost yield. Moringa leaf extract has been identified as one of the natural biostimulants for common bean (Howladar, 2014), wheat (Khursheed et al., 2024), pea (Merwad, 2018), cucumber

(Kakbra, 2024), and sweet pepper (Dlamini et al., 2024). Antioxidants, substances that promote plant growth, and macro- and micronutrients are abundant in melatonin. These nutrients are essential for controlling the physiological, biochemical, and yield parameters of plants under both normal and stressful ecological conditions (Sami et al., 2020).

A typical non-enzymatic antioxidant, ascorbic acid (AsA) has the ability to scavenge ROS and regulates various metabolic processes that are important in both normal and stressful situations. Exogenous application of AsA might dramatically boost endogenous AsA production in wheat tissues. This could be important for the production of certain essential components that would initiate a potent defense against cadmium stress. Moreover, AsA improves physiological and biochemical processes in wheat (Alamri et al., 2018), and barley (Wu and Zhang, 2002). Comparing the advantages and disadvantages of employing pure organic active chemicals to boost yield in the face of metal toxicity is therefore intriguing (Zhang et al., 2019).

Phytohormones are essential for plants' response to metal toxicity (Emamverdian et al., 2020). During metal contact, ethylene specifically regulates several kinds of signaling mechanisms and interactions with minerals and intracellular redox status (Thao et al., 2015). The progress of plants may be impacted by the major effect of metals on intrinsic phytohormonal concentrations (Lucena et al., 2015). Nevertheless, exogenous phytohormones, including gibberellins, ethylene, brassinosteroids, auxins, cytokinins, salicylic acid, and abscisic acid, can moderate metal toxicity (Polle and Schützendübel, 2003). These involve limiting the flow of heavy metals into plant cells, controlling the activity of antioxidant enzymes, and rupturing membranes. Impact of biostimulants on plant health subjected to metal toxicity (Figure 3).

*Figure 3. Impact of biostimulants on plant performance subjected to metal toxicity*

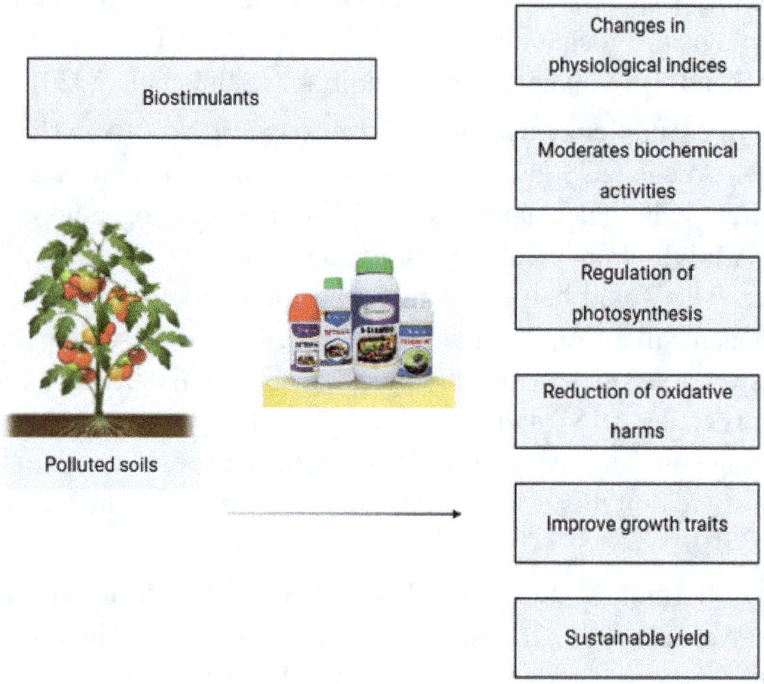


## CONCLUSION

Maintaining the health of the soil and increasing crop yield are always priorities in sustainable agriculture. However, plants must deal with various environmental stressors on a daily basis, which impairs their physiology, growth, and development and lowers their productivity and output. Mutual effects of any stress include the generation of reactive oxygen species, disruption of plant nutritional balance, and disruption of photosynthetic processes and other metabolic cycles. One of the major stressors that the plant should start its defense system against is heavy metal contamination. One such strategy to strengthen the defensive approach to manage and control the general processes of growth and development is the use of biostimulants. Biostimulants use a variety of strategies to combat different biotic and abiotic stressors. To lessen the harmful effects of heavy metals, this chapter aims to provide a brief explanation of biostimulants, their types, and the function of a few key biostimulants. This will establish a foundation for comprehending the many kinds of biostimulants and pave the way for further research into their intricate mechanisms in response to metal toxicity. This chapter will have the overall effect of providing a comprehensive framework for determining the way that plant biostimulants can improve plant resilience to nutrient control, which is common in organic farming, and thereby close the yield gap between organic and conventional farming systems.

## REFERENCES

Adnan, M., Yaseen, T. Hafsa, Khan, M. N., Hayat, K., Wahab, S., Wahid, N., & Kaplan, A. (2025). Use of biostimulants to increase heavy metal tolerance in oilseed crops. In *Oilseed crops under abiotic stress: Mitigation strategies and future perspectives* (pp. 175–197). Springer Nature Singapore.

Ahmad, R., Muhammad, H. M. D., Naz, S., Tiwari, R. K., Lal, M. K., Ahmad, P., & Altaf, M. A. (2024). From stress to success: Silicon and nano-silicon strategies for enhancing fruit yield and quality in cadmium-exposed trees. *South African Journal of Botany, 167*, 441–447. DOI: 10.1016/j.sajb.2024.02.042

Al-Charchafchi, F., & Al-Quadan, F. (2006). Effect of chlorogenic acid on germination and seedling growth, and on the enzymes activity extracted from *Artemisia herba alba* Asso. Part I: Germination and seedling growth. *Dirasat. Pure Sciences, 33*(2), 168–175.

Alamri, S. A., Siddiqui, M. H., Al-Khaishany, M. Y., Nasir Khan, M., Ali, H. M., Alaraidh, I. A., Alsahli, A. A., Al-Rabiah, H., & Mateen, M. (2018). Ascorbic acid improves the tolerance of wheat plants to lead toxicity. *Journal of Plant Interactions, 13*(1), 409–419. DOI: 10.1080/17429145.2018.1491067

Alengebawy, A., Abdelkhalek, S. T., Qureshi, S. R., & Wang, M. Q. (2021). Heavy metals and pesticides toxicity in agricultural soil and plants: Ecological risks and human health implications. *Toxics, 9*(3), 42. DOI: 10.3390/toxics9030042

Alfonso, E., Andlauer, W., Brück, W. M., & Rienth, M. (2025). Whey protein hydrolysates enhance grapevine resilience to abiotic and biotic stresses. *Frontiers in Plant Science, 16*, 1521275. DOI: 10.3389/fpls.2025.1521275

Altaf, M. A., Naz, S., Kumar, R., Sardar, H., Nawaz, M. A., Kumar, A., Lal, P., Ahmad, R., Hayat, F., Wani, M. A., Tiwari, R. K., & Lal, M. K. (2023). Unraveling the mechanisms of cadmium toxicity in horticultural plants: Implications for plant health. *South African Journal of Botany, 163*, 433–442. DOI: 10.1016/j.sajb.2023.10.064

Altaf, M. A., Shahid, R., Naz, S., Ahmad, R., Manzoor, M. A., Alsahli, A. A., Altaf, M. M., & Ahmad, P. (2024). Vanadium toxicity was alleviated by supplementation of silicon in tomato seedlings: Upregulating antioxidative enzymes and glyoxalase system. *Plant Physiology and Biochemistry : PPB, 213*, 108809. DOI: 10.1016/j.plaphy.2024.108809

Anwar, S., Shafiq, F., Nisa, Z. U., Usman, U., Ashraf, M. Y., & Ali, N. (2021). Effect of cadmium stress on seed germination, plant growth and hydrolyzing enzymes activities in mungbean seedlings. *Journal of Seed Science, 43*, e202143042. DOI: 10.1590/2317-1545v43256006

Arikan-Algul, Y., Mutlu-Durak, H., Kutman, U. B., & Yildiz Kutman, B. (2025). Biostimulant extracts obtained from the brown seaweed *Cystoseira barbata* enhance the growth, yield, quality, and nutraceutical value of soil-grown tomato. *Agronomy (Basel), 15*(5), 1138. DOI: 10.3390/agronomy15051138

Basit, F., Khalid, M., Bahadur, A., Bhat, J. A., Ahmad, A., & Ahmad, P. (2025). Alleviation of vanadium toxicity through individual and combined supplementation of spermine and spermidine in soybean: Insights into oxidative stress management, detoxification mechanisms, and root exudates. *South African Journal of Botany, 184*, 468–479. DOI: 10.1016/j.sajb.2025.06.016

Bayata, A. (2020). Assessment, accumulation, toxicity and importance of heavy metals in agricultural soil and living system. *American Journal of Environmental Protection, 9*(6), 121–124. DOI: 10.11648/j.ajep.20200906.11

Bint Tariq, T., & Umer, M. (2025). The potential of seaweed-derived polysaccharides as sustainable biostimulants in agriculture. *International Journal of Biological Macromolecules, 298*, 140009. DOI: 10.1016/j.ijbiomac.2025.140009

Biswas, A., & Pal, S. (2025). Role of phytohormones in developing heavy metal stress resilience in rice: Physiological–molecular adjustments, cross-talk, and prospects. *Journal of Plant Growth Regulation, 44*(9), 4960–4980. Advance online publication. DOI: 10.1007/s00344-025-11757-w

Buss, W., Belt, K., Hein, Z., Putri, A., Zhu, S., Suwandari, H., & Bentley, A. R. (2025). Hurdles to overcome to achieve biostimulant-driven, low chemical input crop production. *Plants, People, Planet.*

Carillo, P., Avice, J. C., Vasconcelos, M. W., Du Jardin, P., & Brown, P. H. (2025). Biostimulants in agriculture. [Adjust article ID if needed]. *Physiologia Plantarum, 177*(1), e70000. DOI: 10.1111/ppl.70046

Carrasco-Castilla, J., Hernández-Álvarez, A. J., Jiménez-Martínez, C., Jacinto-Hernández, C., Alaiz, M., Girón-Calle, J., Vioque, J., & Dávila-Ortiz, G. (2012). Antioxidant and metal chelating activities of peptide fractions from phaseolin and bean protein hydrolysates. *Food Chemistry, 135*(3), 1789–1795. DOI: 10.1016/j.foodchem.2012.06.016

Chai, T. T., Tong, S. R., Law, Y. C., Ismail, N. I. N., Manan, F. A., & Wong, F. C. (2015). Anti-oxidative, metal chelating and radical scavenging effects of protein hydrolysates from blue-spotted stingray. *Tropical Journal of Pharmaceutical Research, 14*(8), 1349–1355. DOI: 10.4314/tjpr.v14i8.5

Choudhary, L. K., & Kumar, A. (2025). Concentration-dependent dual effects of exogenous barium on growth, oxidative stress, and antioxidative enzymes in *Triticum aestivum. Russian Journal of Plant Physiology: a Comprehensive Russian Journal on Modern Phytophysiology, 72*(4), 123. DOI: 10.1134/S1021443724607547

Clemens, S. (2006). Toxic metal accumulation, responses to exposure and mechanisms of tolerance in plants. *Biochimie, 88*(11), 1707–1719. DOI: 10.1016/j.biochi.2006.07.003

Čmiková, N., Galovičová, L., Miškeje, M., Borotová, P., Kluz, M., & Kačániová, M. (2022). Determination of antioxidant, antimicrobial activity, heavy metals and elements content of seaweed extracts. *Plants, 11*(11), 1493. DOI: 10.3390/plants11111493

DalCorso, G., Manara, A., & Furini, A. (2013). An overview of heavy metal challenge in plants: From roots to shoots. *Metallomics : Integrated Biometal Science, 5*(9), 1117–1132. DOI: 10.1039/c3mt00038a

Dalvi, A. A., & Bhalerao, S. A. (2013). Response of plants towards heavy metal toxicity: An overview of avoidance, tolerance and uptake mechanism. *Annals of Plant Sciences*, *2*(9), 362–368.

Derrick, A., Yohana, M. A., Yudong, Z., Li, G., Tan, B., & Zhang, S. (2025). Understanding the detrimental effects of heavy metal pollution in shrimp farming and treatment methods—A review. *Annals of Animal Science*, *25*(1), 35–56. DOI: 10.2478/aoas-2024-0041

Dhalaria, R., Kumar, D., Kumar, H., Nepovimova, E., Kuča, K., Torequl Islam, M., & Verma, R. (2020). Arbuscular mycorrhizal fungi as potential agents in ameliorating heavy metal stress in plants. *Agronomy (Basel)*, *10*(6), 815. DOI: 10.3390/agronomy10060815

Dhiman, S., Khanna, K., Kumar, P., Bhardwaj, T., Devi, K., Sharma, N., Sharma, P., Arora, P., Kapoor, N., Sharma, A., & Bhardwaj, R. (2025). Divulging molecular perspectives of plant defense machinery under heavy metal toxicity. *Journal of Plant Growth Regulation*, *44*(4), 1352–1388. DOI: 10.1007/s00344-023-11138-1

Dlamini, S. W., Bertling, I., & Ngcobo, B. L. (2024). Enhancing growth, yield and physiological quality of sweet pepper fruit through application of fertiliser and moringa leaf powder (MLP). *Acta Agriculturæ Scandinavica. Section B, Soil and Plant Science*, *74*(1), 2396973. DOI: 10.1080/09064710.2024.2396973

Einhardt, A. M., Ferreira, S., & Rodrigues, F. Á. (2021). Biochemical and physiological responses of soybean [*Glycine max* (L.) Merrill] to nickel toxicity. *Bragantia, 80*, Article e1721.

El Khattabi, O., El Hasnaoui, S., Toura, M., Henkrar, F., Collin, B., Levard, C., Colin, F., Merghoub, N., Smouni, A., & Fahr, M. (2023). Seaweed extracts as promising biostimulants for enhancing lead tolerance and accumulation in tomato (*Solanum lycopersicum*). *Journal of Applied Phycology*, *35*(1), 459–469. DOI: 10.1007/s10811-022-02849-1

El-Mogy, M. M., Jan, M. F., Naz, I., Ahmad, I., Ahmad, R., & Altaf, M. T. (2025). Nanoparticle innovations for mitigating metal toxicity in plants. *Phyton*, *94*(3), 623–640. DOI: 10.32604/phyton.2025.063763

Emamverdian, A., Ding, Y., & Xie, Y. (2020). The role of new members of phytohormones in plant amelioration under abiotic stress with an emphasis on heavy metals. *Polish Journal of Environmental Studies, 29*(2), 1009–1020. DOI: 10.15244/pjoes/108687

Ghuge, S. A., Nikalje, G. C., Kadam, U. S., Suprasanna, P., & Hong, J. C. (2023). Comprehensive mechanisms of heavy metal toxicity in plants, detoxification, and remediation. *Journal of Hazardous Materials, 450*, 131039. DOI: 10.1016/j.jhazmat.2023.131039

Haghighi, M., Kafi, M., Fang, P., & Gui-Xiao, L. (2010). Humic acid decreased hazardous of cadmium toxicity on lettuce (*Lactuca sativa* L.). *Vegetable Crops Research Bulletin, 72*, 49.

Haiying, T., Shubin, W., Ying, L., Ying, S., & Yasser, S. M. (2022). Biochar: A promising soil amendment to mitigate heavy metals toxicity in plants. *Notulae Botanicae Horti Agrobotanici Cluj-Napoca, 50*(3), 12778. DOI: 10.15835/nbha50312778

Hasan, M., & Naushin, F. (2026). Plant responses to heavy metals: Mechanisms and adaptations. In *Optimizing plant health under abiotic-stress environments* (pp. 1–28). CRC Press.

Hasan, M., Naushin, F., & Khan, F. A. (2025). Soil symbionts in alleviation of heavy metals-laden sewage sludge-impacts in legumes: Implications for phytoremediation. *Discover Plants, 2*(1), 103. DOI: 10.1007/s44372-025-00192-6

Howladar, S. M. (2014). A novel *Moringa oleifera* leaf extract can mitigate the stress effects of salinity and cadmium in bean (*Phaseolus vulgaris* L.) plants. *Ecotoxicology and Environmental Safety, 100*, 69–75. DOI: 10.1016/j.ecoenv.2013.11.022

Huang, L., Lee, J. Y., Park, Y. K., & Lee, J. (2025). Heavy metals in seaweed: Implications for health benefits, risks, and safety regulations. *Journal of Agriculture and Food Research, 21*, 101830. DOI: 10.1016/j.jafr.2025.101830

Kakbra, R. F. (2024). Effect of seaweed, moringa leaf extract and biofertilizer on growth, yield and fruit quality of cucumber (*Cucumis sativus* L.) under greenhouse condition. arXiv. https://arxiv.org/abs/2403.17984

Kalyabina, V. P., Esimbekova, E. N., Kopylova, K. V., & Kratasyuk, V. A. (2021). Pesticides: Formulants, distribution pathways and effects on human health—A review. *Toxicology Reports*, *8*, 1179–1192. DOI: 10.1016/j.toxrep.2021.06.004

Kaur, R., Choudhary, D., Bali, S., Bandral, S. S., Singh, V., Ahmad, M. A., Rani, N., Singh, T. G., & Chandrasekaran, B. (2024). Pesticides: An alarming detrimental to health and environment. *The Science of the Total Environment*, *915*, 170113. DOI: 10.1016/j.scitotenv.2024.170113

Khan, N., Ryu, K. Y., Choi, J. Y., Nho, E. Y., Habte, G., Choi, H., Kim, M. H., Park, K. S., & Kim, K. S. (2015). Determination of toxic heavy metals and speciation of arsenic in seaweeds from South Korea. *Food Chemistry*, *169*, 464–470. DOI: 10.1016/j.foodchem.2014.08.020

Khoulati, A., Ouahhoud, S., Taibi, M., Ezrari, S., Mamri, S., Merah, O., Hakkou, A., Addi, M., Maleb, A., & Saalaoui, E. (2025). Harnessing biostimulants for sustainable agriculture: Innovations, challenges, and future prospects. *Discover Agriculture, 3*(1), Article 56.

Khursheed, M. H. U. R., Shahbaz, M., Ramzan, T., Haider, A., Maqsood, M. F., Khan, A., Zulfiqar, U., Jamil, M., Hussain, S., Al-Ghamdi, A. A., & Rizwana, H. (2024). Enhancing wheat tolerance to cadmium stress through moringa leaf extract foliar application. *Scientifica*, *2024*(1), 2919557. DOI: 10.1155/2024/2919557

Lanrewaju, A. A., & Fadiji, A. E. (2025). Microbial biostimulants: Boosting plant performance and stress tolerance. In *Microbial allies* (pp. 147–176). Springer. DOI: 10.1007/978-3-031-90530-8_8

Liu, J., Gai, L., & Zong, H. (2021). Foliage application of chitosan alleviates the adverse effects of cadmium stress in wheat seedlings (*Triticum aestivum* L.). *Plant Physiology and Biochemistry : PPB*, *164*, 115–121. DOI: 10.1016/j.plaphy.2021.04.038

Loague, K., & Corwin, D. L. (2006). Point and nonpoint source pollution. In *Encyclopedia of hydrological sciences*.

Lucena, C., Romera, F. J., García, M. J., Alcántara, E., & Pérez-Vicente, R. (2015). Ethylene participates in the regulation of Fe deficiency responses in Strategy I plants and in rice. *Frontiers in Plant Science*, *6*, 1056. DOI: 10.3389/fpls.2015.01056

Manna, I., Nath, S., Mandal, P., & Paul, S. (2025). Effect of cerium oxide and silicon dioxide nanoparticles in combating heavy metal stress in plants. *The Nucleus : an International Journal of Cytology and Allied Topics*. Advance online publication. DOI: 10.1007/s13237-025-00582-x

Manzoor, M., Korolev, K., & Ahmad, R. (2025). Zinc oxide nanoparticles: Abiotic stress tolerance in fruit crops focusing on sustainable production. *Phyton*, *94*(5), 1401–1418. DOI: 10.32604/phyton.2025.063930

Merwad, A. R. M. (2018). Using *Moringa oleifera* extract as biostimulant enhancing the growth, yield and nutrients accumulation of pea plants. *Journal of Plant Nutrition*, *41*(4), 425–431. DOI: 10.1080/01904167.2017.1384012

Muhammad, H. M. D., Naz, S., Bibi, Z., Ahmed, M. A., Stanciu, A. S., & Ahmad, R. (2025). Nanoparticles and phytohormonal synergy in plants: Sustainable agriculture approach. *Phyton*, *94*(9), 2631–2648. DOI: 10.32604/phyton.2025.069474

Muhammad, H. M. D., Naz, S., Lal, M. K., Tiwari, R. K., Ahmad, R., Nawaz, M. A., Das, R., & Altaf, M. A. (2024). Melatonin in business with abiotic stresses in vegetable crops. *Scientia Horticulturae*, *324*, 112594. DOI: 10.1016/j.scienta.2023.112594

Naz, S., Anjum, M. A., Sadiq, B., Ahmad, R., Altaf, M. A., El-Sheikh, M. A., & Shakoor, A. (2022). Purification of sewage wastewater though sand column filter for lessening of heavy metals accumulation in lettuce, carrot, and cauliflower. *Water (Basel)*, *14*(22), 3770. DOI: 10.3390/w14223770

Ochoa, L., Shrivastava, M., Srivastava, S., Cota-Ruiz, K., Zhao, L., White, J. C., Hernandez-Viezcas, J. A., & Gardea-Torresdey, J. L. (2025). Nanomaterials for managing abiotic and biotic stress in the soil–plant system for sustainable agriculture. *Environmental Science. Nano*, *12*(2), 1037–1058. DOI: 10.1039/D4EN00789A

Oliveira, D. J. B., Lavres, J., Kopittke, P. M., Chaney, R. L., Harris, H. H., Erskine, P. D., Howard, D. L., dos Reis, A. R., & van der Ent, A. (2024). Unravelling the fate of foliar-applied nickel in soybean: A comprehensive investigation. *Plant and Soil*, *502*(1), 537–556. DOI: 10.1007/s11104-024-06567-0

Ondrasek, G., Rengel, Z., & Romic, D. (2018). Humic acids decrease uptake and distribution of trace metals, but not the growth of radish exposed to cadmium toxicity. *Ecotoxicology and Environmental Safety*, *151*, 55–61. DOI: 10.1016/j.ecoenv.2017.12.055

Özkay, F., Kıran, S., Kuşvuran, Ş., & Ellialtıoğlu, Ş. Ş. (2016). The effects of humic acid applications on heavy metal stress in lettuce. *Turkish Journal of Agriculture - Food Science and Technology*, *4*(6), 431–437.

Pasković, I., Popović, L., Pongrac, P., Polić Pasković, M., Kos, T., Jovanov, P., & Franić, M. (2024). Protein hydrolysates—Production, effects on plant metabolism, and use in agriculture. *Horticulturae*, *10*(10), 1041. DOI: 10.3390/horticulturae10101041

Perelomov, L. V., Sarkar, B., Sizova, O. I., Chilachava, K. B., Shvikin, A. Y., Perelomova, I. V., & Atroshchenko, Y. M. (2018). Zinc and lead detoxifying abilities of humic substances relevant to environmental bacterial species. *Ecotoxicology and Environmental Safety*, *151*, 178–183. DOI: 10.1016/j.ecoenv.2018.01.018

Polle, A., & Schützendübel, A. (2003). Heavy metal signalling in plants: Linking cellular and organismic responses. In *Plant responses to abiotic stress* (pp. 187–215). Springer. DOI: 10.1007/978-3-540-39402-0_8

Raffa, C. M., & Chiampo, F. (2021). Bioremediation of agricultural soils polluted with pesticides: A review. *Bioengineering (Basel, Switzerland)*, *8*(7), 92. DOI: 10.3390/bioengineering8070092

Rao, M. J., Duan, M., Zhou, C., Jiao, J., Cheng, P., Yang, L., Wei, W., Shen, Q., Ji, P., Yang, Y., Conteh, O., Yan, D., Yuan, H., Rauf, A., Ai, J., & Zheng, B. (2025). Antioxidant defense system in plants: Reactive oxygen species production, signaling, and scavenging during abiotic stress-induced oxidative damage. *Horticulturae*, *11*(5), 477. DOI: 10.3390/horticulturae11050477

Rasheed, R., Ashraf, M. A., Arshad, A., Iqbal, M., & Hussain, I. (2020). Interactive effects of chitosan and cadmium on growth, secondary metabolism, oxidative defense, and element uptake in pea (*Pisum sativum* L.). *Arabian Journal of Geosciences*, *13*(17), 847. DOI: 10.1007/s12517-020-05871-0

Rong, Q., Zhong, K., Huang, H., Li, C., Zhang, C., & Nong, X. (2020). Humic acid reduces the available cadmium, copper, lead, and zinc in soil and their uptake by tobacco. *Applied Sciences (Basel, Switzerland)*, *10*(3), 1077. DOI: 10.3390/app10031077

Rucińska-Sobkowiak, R. (2016). Water relations in plants subjected to heavy metal stresses. *Acta Physiologiae Plantarum*, *38*(11), 257. DOI: 10.1007/s11738-016-2277-5

Sah, S. K., Mhetre, A. G., Kumari, S., Gadi, Y., Prajapat, M., Kashyap, M., Adhikari, S., & Kumar, A. (2025). Emerging techniques of soil remediation: Nanotechnology and bioremediation approaches. *International Journal of Advanced Biochemistry Research*, *9*(1), 488–504. DOI: 10.33545/26174693.2025.v9.i1f.3560

Sami, A., Shah, F. A., Abdullah, M., Zhou, X., Yan, Y., Zhu, Z., & Zhou, K. (2020). Melatonin mitigates cadmium and aluminium toxicity through modulation of antioxidant potential in *Brassica napus* L. *Plant Biology*, *22*(4), 679–690. DOI: 10.1111/plb.13093

Sardar, K., Ali, S., Hameed, S., Afzal, S., Fatima, S., Shakoor, M. B., Bharwana, S. A., & Tauqeer, H. M. (2013). Heavy metals contamination and what are the impacts on living organisms. *Greener Journal of Environmental Management and Public Safety*, *2*(4), 172–179. DOI: 10.15580/GJEMPS.2013.4.060413652

Schutzendubel, A., & Polle, A. (2002). Plant responses to abiotic stresses: Heavy metal-induced oxidative stress and protection by mycorrhization. *Journal of Experimental Botany*, *53*(372), 1351–1365. DOI: 10.1093/jexbot/53.372.1351

Shaikh, A., Jamla, M., Sharma, L., Palawat, B., Pathak, S., Shekhawat, G. S., Tripathi, M., & Kumar, V. (2025). Multifaceted roles of biogenic nanoparticles for conferring heavy metal tolerance in plants. *Journal of Crop Health*, *77*(3), 82. DOI: 10.1007/s10343-025-01147-3

Shar, A. G., Zhang, L., Lu, A., Ahmad, M., Saqib, M., Hussain, S., Zulfiqar, U., Wang, P., Zhang, L., & Rahimi, M. (2025). Unlocking biochar's potential: Innovative strategies for sustainable remediation of heavy metal stress in tobacco plants. *Scientifica*, *2025*(1), 6302968. DOI: 10.1155/sci5/6302968

Sharma, A., Dubey, S., Singh, K., Mittal, R., Quille, P., & Rajauria, G. (2025). Innovative processing and industrial applications of seaweed. *Phycology*, *5*(1), 10. DOI: 10.3390/phycology5010010

Sharma, S. S., & Dietz, K. J. (2009). The relationship between metal toxicity and cellular redox imbalance. *Trends in Plant Science*, *14*(1), 43–50. DOI: 10.1016/j.tplants.2008.10.007

Shivappa, S., Amritha, K. P., Nayak, S., Chandrashekar, H. K., Thorat, S. A., Kaniyassery, A., Govender, N., Thiruvengadam, M., & Muthusamy, A. (2025). Integration of physio-biochemical, biological and molecular approaches to improve heavy metal tolerance in plants. *3 Biotech, 15*(4), 1–28.

Song, C., Bibi, Z., Ali, S., & Ahmad, R. (2026). Silicon mitigates aluminum toxicity in lettuce by improving photosynthetic pigments and antioxidant defense system. *Biometals*. Advance online publication. DOI: 10.1007/s10534-026-00819-5

Sreelatha, L., Ambili, A. L., Sreedevi, S. C., & Achuthavarier, D. (2025). Metallothioneins: An unraveling insight into remediation strategies of plant defense mechanisms. *Environmental Science and Pollution Research International*, *32*(2), 405–427. DOI: 10.1007/s11356-024-35790-6

Srikanth, H., Kumari, K., & Meena, R. (2022). Contamination of groundwater with heavy metals and its impact on soil and plants—A review. *Ecology. Environmental Conservation*, *28*, S331–S335.

Srivastava, A. K., Kumari, S., Singh, R. P., Khan, M., Mishra, P., & Xie, X. (2025). Harnessing the interplay of protein posttranslational modifications: Enhancing plant resilience to heavy metal toxicity. *Microbiological Research*, •••, 128112.

Medvedkina, Y. A., & Khodochenko, A. V. (2020, October). Renewable energy and their impact on environmental pollution in the context of globalization. In 2020 International Multi-Conference on Industrial Engineering and Modern Technologies (FarEastCon) (pp. 1-4). IEEE.

Subin, M. P., & Francis, S. (2013). Phytotoxic effects of cadmium on seed germination, early seedling growth and antioxidant enzyme activities in *Cucurbita maxima* Duchesne. *International Research Journal of Biological Sciences*, *2*(9), 40–47.

Suman, S. N., Pradhan, B., Kumar, V., Singh, H., Nath, D., Kamat, D. N., & Kumar, M. (2025). Exploring the role of biostimulants in sustainable agriculture. *Current Science*, *128*(1), 25–30. DOI: 10.18520/cs/v128/i1/25-30

Tahir, R., Afzal, F., Jamil, H., Razzaq, M., & Khan, M. S. (2024). Physiological impacts of pesticidal contamination: Challenge to sustainable agriculture and biodegradation methods. *Advances in Agriculture and Animal Science*, *40*(1), 24–37. DOI: 10.47432/2024.40.1.4

Thakur, S., Walia, S., Thakur, B., Ghosh, A., & Kumar, R. (2025). Unleashing the potential of seaweed biostimulants—A comparative evaluation for enhancing saffron (*Crocus sativus* L.) yield with different corm sizes in the Western Himalayas. *BMC Plant Biology*, *25*(1), 946. DOI: 10.1186/s12870-025-06955-3

Thao, N. P., Khan, M. I. R., Thu, N. B. A., Hoang, X. L. T., Asgher, M., Khan, N. A., & Tran, L. S. P. (2015). Role of ethylene and its cross talk with other signaling molecules in plant responses to heavy metal stress. *Plant Physiology*, *169*(1), 73–84. DOI: 10.1104/pp.15.00663

Ullah, N., Ur Rehman, M., Ahmad, B., Ali, I., Younas, M., Aslam, M. S., Rahman, A. U., Taheri, E., Fatehizadeh, A., & Rezakazemi, M. (2022). Assessment of heavy metals accumulation in agricultural soil, vegetables and associated health risks. *PLoS One*, *17*(6), e0267719. DOI: 10.1371/journal.pone.0267719

Uruç Parlak, K. (2016). Effect of nickel on growth and biochemical characteristics of wheat (*Triticum aestivum* L.) seedlings. *NJAS Wageningen Journal of Life Sciences*, *76*(1), 1–5. DOI: 10.1016/j.njas.2012.07.001

Wang, M., Chen, L., Wu, L., Zhang, L., Xie, H., & Shen, Z. (2022). Review of nonpoint source pollution models: Current status and future direction. *Water (Basel)*, *14*(20), 3217. DOI: 10.3390/w14203217

Wu, F., & Zhang, G. (2002). Alleviation of cadmium-toxicity by application of zinc and ascorbic acid in barley. *Journal of Plant Nutrition*, *25*(12), 2745–2761. DOI: 10.1081/PLN-120015536

Yadav, A., Hasan, M., Kour, S., Chandra, R., Singh, V., Noor, I. M., Yahya, M. Z. A., Singh, H. M., & Tripathi, A. (2025). Heavy metal contamination in plants: Sources, monitoring, and data-driven insights. *Macromolecular Symposia*, *414*(1), 2400239. DOI: 10.1002/masy.202400239

Zhang, K., Wang, G., Bao, M., Wang, L., & Xie, X. (2019). Exogenous application of ascorbic acid mitigates cadmium toxicity and uptake in maize (*Zea mays* L.). *Environmental Science and Pollution Research International*, *26*(19), 19261–19271. DOI: 10.1007/s11356-019-05265-0

# Chapter 4
# Nitric Oxide Reduces the Adverse Effects of Metal Toxicity in Plants

**Saqib Ali**
http://orcid.org/0000-0002-6338-4385
*The University of Agriculture, Dera Ismail Khan, Pakistan*

## ABSTRACT

*Nitric oxide (NO) is a redox related signaling molecules that is thought to be important regulators involved in plant growth and various developmental processes as well as the way plants react to abiotic stress conditions. Metal toxicity causes reactive oxygen species (ROS) to be produced, which brings plants under oxidative harm. Plants use various ROS-scavenging systems, including signaling molecules, non-enzymatic antioxidants and antioxidative enzymes, to reduce the negative moments of metal toxicity. NO has high potential to mitigate the adverse effects of metal toxicity by regulation of ROS overproduction. The way in which nitric oxide triggers the proper set of reactions to metal toxicity requires a comprehension of its functions. In this chapter concentrate on the most recent information regarding nitric oxide's mitigating contribution to metal toxicity,encompassing everything compared to metal toxicity,build-up to the antioxidant response to associated gene and protein regulation.*

DOI: 10.4018/979-8-3693-9899-9.ch004

## INTRODUCTION

Nitric oxide is thought to be a common transmitter of signals that controls a number of plant-related workflows, including stomatal closing, seed germination, and root growth (Liu and Zhang, 2021). Nitric oxide is typically the primary enzymes involved in the synthesis of endogenous nitric oxide and nitrate reductase. Nitric oxide plays a significant role in the response system to abiotic stressors (Zhang et al., 2023). Metal toxicity is one of the agricultural pollutions that can result from prolonged consumption of phosphatic nutrients, sewage solids, industrial debris, and irrigation with brackish water, which can pollute crops on farms (Rashid et al., 2023). Excessive production of Reactive oxygen species (ROS) and rupture of membranes are the primary causes of the harmful impact of metal toxicity on plants, which in turn causes oxidative injury in stressed crops (Rashid et al., 2023). It has been evaluated that plants actively generate ROS in reaction to stress conditions like metal stress. Plants had different ROS-scavenging systems that primarily antioxidative enzymes and non-enzymatic enzymes involved in the reduction of harmful effects stresses (Melebary, 2023). Likewise, ability of plants to tolerate stresses enhanced with activation of antioxidant enzymes that eliminate free radicals from oxygen (Tak et al., 2012). Generally, it has been explored that nitric oxide regulates iron buildup, ROS, and antioxidants under metal stress conditions (Xiong et al., 2010).

Nitric oxide regulates the transcription level linked to associated genes as well as the synthesis of proteins that improved plant tolerance. A nitric oxide donor is sodium nitroprusside that may boost metal toxicity with the activation of gene transcriptions relevant to stress proteins (Saxena and Shekhawat, 2013). Abscisic acid-induced the acceleration of the mitogen-activated protein kinase (MAPK) flow, which contributes in defense system in maize leaves, was facilitated by nitric oxide produced by hydrogen peroxide ($H_2O_2$) under metal toxicity (Liu et al., 2020). Moreover, the interaction among nitric oxide as well as various other phytohormones that is now being documented throughout the early stages of plant life, including seed dormancy as well as germination, hypocotyl expansion, and root growth (Okant and Kaya, 2019). Protein S-nitrosylation, a crucial posttranslational alteration as well as a crucial mechanism by which nitric oxide controls cellular redox signaling under metal stress (Mata-Pérez et al., 2023). Moreover, it is yet to be investigated and compiled if this interaction includes ROS, antioxidants, the transcription of linked genes, and protein action (Seifried et al., 2007). Consequently, recent research illustrated the relieving activities of nitric oxide in metal toxicity in order to more accurately comprehend the interplay involving nitric oxide and metal toxicity (Wei et al., 2020). Explicitly, by controlling the amount of iron metal and triggering antioxidant systems, nitric oxide may reduce metal toxicity (Saxena and Shekhawat, 2013). Plant responses to metal toxicity involved the regulation of homologous protein-coding genes (Sharma et al., 2020).

Metal toxicity significantly reducing the productivity of agriculture crops globally. The mitigation of adverse effects of metal toxicity is necessary for sustainable farming practices. Therefore, different management practices are involved in the mitigation of metal toxicity focusing on growth, yield, and quality. Hence, current chapter fully explore the mitigation of metal toxicity with supplementation of nitric oxide.

## ADVERSITIES OCCUR FROM METAL TOXICITY IN PLANTS

One of the primary soil pollutants limiting crop productivity worldwide is metal toxicity. A decrease in yield results from plants cultivating in polluted soils exhibiting stunted growth and irregular developmental processes (Küpper and Andresen, 2016). The maximum soil concentrations of metals cause high absorption and affect various metabolic processes (Clemens, 2006). Metals in excess can change the shape and structure of roots and shoots as they progress (Zhang et al., 2017). Metal toxicity greatly decreased the plant growth and depress the various developmental processes of several plant species eggplant (Körpe and Aras, 2011), cucumber (Wang et al., 2020), mulberry (Huihui et al., 2020), peppers (Altaf et al., 2022), mango (Širić et al., 2022), tomato (Ran et al., 2023), chilli (Mushtaq et al., 2024). Plants have developed efficient ways to induce the synthesis of metals carriers via signaling routes, as well as have intricate biological and biological defense systems, likewise conserving and delivering metals, to defend truly regarding metal toxicity (Mandal et al., 2022). To deal with metal toxicities, plants either prevent retaining metals in the soil and activate the eliminating process, which involves generating antioxidant substances for the eliminating of ROS. Cadmium significantly reduced growth characteristics and triggered oxidative injury through boosting electrolyte leakage, malondialdehyde, and proline level in pea seedlings (Hoque et al., 2021). The activation of antioxidant defense system improves metal tolerance in plants (Table 1).

*Table 1. Plant response to activation of antioxidant defense system under metal toxicity*

| Metal name | Crop name | Adverse effects | Antioxidants and signaling molecules | Plants response | Reference |
|---|---|---|---|---|---|
| Arsenic | Rice | Reduced biomass, leaf necrosis | APX, GR, DHAR, MDHAR, and ASA | Redox balance restoration and reduced oxidative stress | Singh et al. (2009) |
| Lead | Maize | Photosynthesis disruption | SOD, POD, CAT, GSH, and ASA | Detoxification and cell membrane protection | Okant and Kaya (2019) |
| Cadmium | Tomato | Growth inhibition, chlorosis, root damage | Phenolics content, flavonoids, SOD, CAT, APX, GR activation. | Enhanced ROS scavenging and partial tolerance | Ahmad et al. (2018) |
| Lead and cadmium | Pepper | Photosynthesis inhibition, protein oxidation | SOD, POD, APX, AsA, and GSH | Metal detoxification, reduction of oxidative injury | Kaya et al. (2019) |
| Mercury | Soybean | Inhibited enzyme activity, chlorosis | SOD, CAT, GST, and GSH | Antioxidant upregulation, improved tolerance | Ahmad et al. (2021) |
| Zinc | Wheat | Nutrient imbalance, membrane damage | SOD, CAT, APX, GR, and GSH | Induces tolerance at high but non-toxic concentrations | Buet et al. (2014) |
| Nickel | Wheat | Reduced chlorophyll, DNA damage | SOD, CAT, APX GSH, proline | Maintains redox homeostasis, supports metabolic function | Siddiqui et al. (2013) |

## NITRIC OXIDE REGULATES PHOTOSYNTHESIS APPARATUS IN PLANTS

Nitric oxide is essential for signaling the manner in which plants react to various kinds of abiotic stressors, such as metal toxicity (Meng et al., 2022). Nitric oxide has several physiological functions under metal toxicity, when photosynthesis is extremely vulnerable to harm, it has become an important regulator of the photosynthesis mechanism. Metal toxicity is producing irregular ROS, altering chloroplast, decreasing chlorophyll content, as well as disrupting the photosynthesis processes (Kumar et al., 2022). However, applying nitric oxide, has been demonstrated to control and lessen these adverse effects, protecting or even regaining photosynthetic performance in distressed plants. Metal toxicity causes harm through interfering with the strength of the thylakoid membrane, altering the ions movement, and preventing photosystems I and II from functioning (Mahjoubi et al., 2025). As a result, there is an important reduction in net photosynthesis, transpiration, chlorophyll content, and carbon fixation level. Nitric oxide regulates preserving redox balance, neutralizing ROS, and strengthening antioxidant defense mechanisms (Ghorbani et al., 2021). Numerous studies have shown that by boosting the functioning of genes that govern the pigment formation and shielding chlorophyll contents via oxidative harm, intrinsic nitric oxide supporters like sodium nitroprusside may reduce the hindering impact of metallic substances on chlorophyll metabolism (Khan et al., 2022; Verma and Prasad, 2021; Basit et al., 2024).

*Figure 1. The activation of plant defense system under metal toxicity in plants*

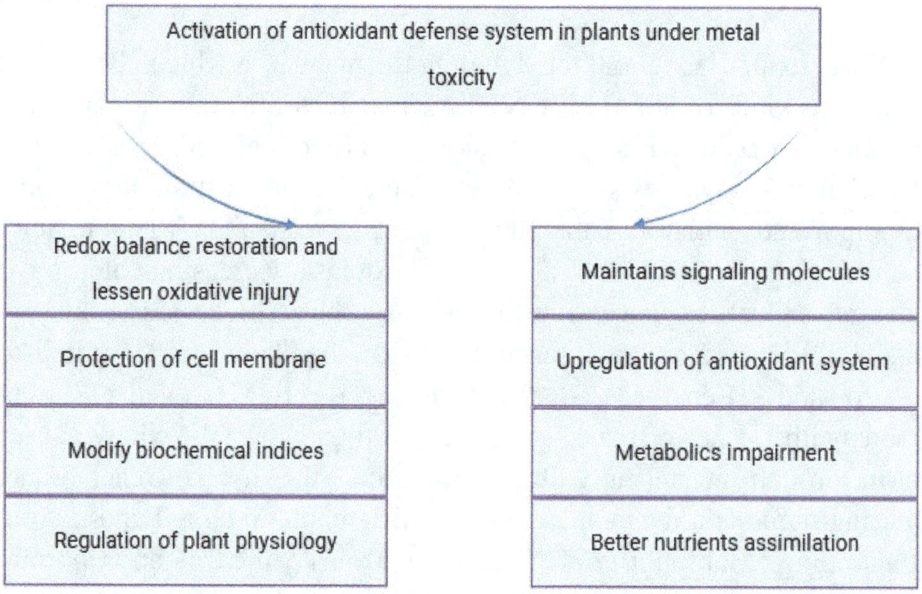


Nitric oxide provides protection by maintaining chloroplasts which further improves photosynthesis (Figure 1). Nitric oxide inhibits the ultrastructural changes in chloroplasts that are usually brought through prolonged exposure to toxic substances (Khan et al., 2022). Furthermore, nitric oxide maintains $CO_2$ acquisition under metal toxicity by regulating the actions of important enzymes throughout the Calvin cycle (Verma and Prasad, 2021). Moreover, it improves PSII effectiveness and strengthens the light-harvesting multifaceted, as evidenced by an increased quantum output (Ahmad et al., 2021). Likewise, nitric oxide functions as a messenger molecule that triggers the functioning of genes vulnerable to stress, especially those linked to enzymes that protect against damage (Imran et al., 2023). This minimizes the oxidation of lipids and cellular harm by lowering the amount of ROS in the chloroplasts. Furthermore, nitric oxide regulates the ascorbate-glutathione process, which is essential for eliminating $H_2O_2$ generated during ions movement throughout photosynthetic processes (Basit et al., 2022). Nitric oxide had potential to encourage the performance of the photosynthetic mechanism by boosting the activation of antioxidant potential of plants subjected to metal toxicity (Ahmad et al., 2021).

Nitric oxide interplay with various signaling molecules and phytohormones regulates photosynthetic activity of plants against metal toxicity. The synergy between phytohormones and signaling molecules improves stress tolerance in metal subjected plants (Singhal et al., 2021). For instance, nitric oxide encourages the generation of metallothionein and heat shock proteins that contributes in the detoxification of metallic ions, resulting in regulation of photosynthetic system (Sami et al., 2018). For nitric oxide application mainly based on its applied doses, selection of plants, and severity of stress. Nitric oxide application at minimal doses can regulate signaling molecules, but it can worsen oxidative harm and become toxic at higher concentrations (Freschi, 2013). Moreover, exogenous nitric oxide regulated the level of nitrate reductase to attain defensive system of plants under metal toxicity (Gill et al., 2013).

Nitric oxide regulates photosynthesis by scavenging of toxic elements under metal toxicity. It increases photosystem efficiency, boosts antioxidant defenses, preserves chloroplast arrangement, and increases chlorophyll content (Mishra et al., 2021). During metal toxicity, these processes work together to restore metabolism and growth in plants. acquiring insight into NO-mediated regulatory systems can help develop agronomic approaches that increase plant resistance to metal pollution (Arif et al., 2021). Hence, nitric oxide is effective for the regulation of photosynthesis occur in plants subjected to metal toxicity (Table 2).

*Table 2. Impact of nitric oxide on the regulation of photosynthesis in plants*

| **Crop name** | **Metal name** | **Photosynthetic pigments** | **Nitric oxide impact** | **Reference** |
|---|---|---|---|---|
| Bamboo plants | Lead and cadmium toxicity | Chlorophyll Content | It regulates chlorophyll content | Emamverdian et al. (2021) |
| Pepper | Chromium | Photosystem II | Boosts Fv/Fm ratio as well as quantum yield | Kaya et al. (2022) |
| Tomato | Nickel | Stomatal conductance | Adjusts stomatal opening and closing | Qin et al. (2022) |

continued on following page

*Table 2. Continued*

| Crop name | Metal name | Photosynthetic pigments | Nitric oxide impact | Reference |
|---|---|---|---|---|
| Eggplant | Nickel | Chloroplast ultrastructure | Sustains intact chloroplast structure under metal toxicity | Soliman et al. (2019) |
| Wheat | Cadmium | Non-photochemical quenching | Moderates NPQ to diminish photo-damage | Nawaz et al. (2024) |
| Wheat | Zinc | Carotenoids and pigments firmness | Alleviates carotenoids and photopigments | Duan et al. (2015) |
| Maize | Cadmium | Photopigments firmness | Diminish oxidative impairment | Liu et al. (2020) |
| Maize | Cadmium | Interaction with many other endogenous hormones | Crosstalk with other phytohormones to regulate photosynthetic purpose | Lee et al. (2025) |

## CROSSTALK OF METAL TOXICITY AND ACTIVATION OF ANTIOXIDANT SYSTEM IN PLANTS

*Figure 2. The impact of nitric oxide on the regulation of photosynthetic pigments under metal toxicity*

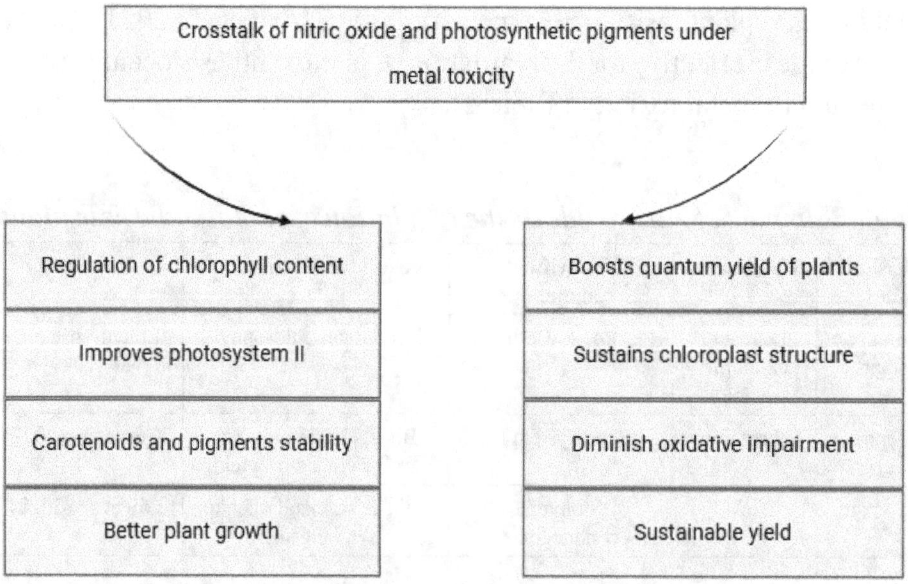

The modulation of antioxidants defense system is impaired due to metal toxicity and illustrated adverse effects on growth and numerous developmental processes occur in plants (Figure 2). Oxidation and cell death result from the interference of metallic substances and elevated levels of critical metals with disruption in physiological as well as biochemical reactions (Raza et al., 2021). The activation and modulation of system of antioxidants are directly related to its capacity to cope in these harsh conditions. To preserve redox homeostasis and minimize damage, intricate signaling paths as well as dynamic metabolic reactions are involved in the interaction between metal toxicity and the antioxidant system (Nahar et al., 2022). The main way that metal poisoning causes injury is by producing too many free radicals and toxic substances especially ROS. Metal-induced impairment of enzyme function, hindering of antioxidant reactions, and intervention with ionic movement in chloroplasts and mitochondria cause ROS accumulation (Raza et al., 2022). Low concentrations of ROS serve as mediators in networks linked to growth and stress, but when they grow up unregulated, may cause membrane problems, oxidation of protein, and lipids (Zhang et al., 2024).

Plants use an intricately synchronized system of antioxidants that consists of both enzymatic as well as non-enzymatic aspects to combat oxidative damage. These are significantly contributing in the scavenging of toxic ROS (Dash et al., 2025). Plants activate their defense system under stress to regulate the overgeneration of ROS because ROS enhanced the oxidative harms (Ghuge et al., 2023). Moreover, membrane leakage, and lipid peroxidation are majorly occurred due to excess toxic ROS (Angulo-Bejarano et al., 2021). However, signaling molecules also contribute in the determination of stress level in plants (Riyazuddin et al., 2021). Hence, exogenous supplementation of nitric oxide is effective for the regulation of toxic ROS. Plant defense system also activated with exogenous supplementation of nitric oxide (Liu et al., 2023).

The main enzymatic antioxidants break down $H_2O_2$ into oxygen as well as water, and different peroxidases use inhibitors to neutralize $H2O_2$. Monodehydroascorbate reductase (MDHAR) as well as glutathione reductase (GR), which aid in the regeneration of ascorbic acid (AsA) and reduced glutathione (GSH), respectively, are essential necessary elements (Astier et al., 2018). In addition to neutralizing ROS, non-enzymatic activities like GSH, AsA, tocopherols, phenolics, and flavonoids are crucial for maintaining cell components. Moreover, GSH plays a crucial role in metal elimination by forming metal-chelate bonds and contributing to the production of phytochelatins, that bind and retain metals towards vacuoles, thereby lowering their cytosolic toxic effects (Gupta et al., 2022).

The kind of metal, level, kind of plant, and phase of growth are some of the variables that affect the interaction between metal stress as well as antioxidant response. As an illustration, it is well known that cadmium and lead significantly lower photosynthesis as well as chlorophyll levels, which leads to a large increase in enzymatic activity (Krasylenko et al., 2010). Indicating an increased robust redox-buffering capability, researches demonstrated that tolerant plant cultivars frequently sustain stronger basal as well as triggered levels of antioxidants than sensitive cultivars (Castello et al., 2019). The signaling molecules promote the triggering of antioxidant reactions under metal toxicity. These compounds combine endogenous defense mechanisms in extrinsic metal signals by modifying the gene expression of antioxidant enzymes and other protective proteins (Buet et al., 2019). Transcription factors (MYB, NAC, WRKY, and bZIP) are also essential for coordinating the antioxidant apparatus in response to metal-induced ROS (Graziano and Lamattina, 2005).

The regulation of signaling network is the indication of stress tolerance in plants under metal toxicity (Hashem et al., 2018). The higher concentration of ROS was due to metal toxicity and its regulation is also the indication of stress tolerance in plants (Begara-Morales et al., 2018).

The crosstalk of metal toxicity and activation of defense system is much considered as toxic for plants overall performance. The signal transduction through gene factors, phytohormones, and the triggering of antioxidants contributed in the elimination of ROS in plant cell compartments (Verma et al., 2020). Different management approaches can mitigate the metal toxicity. Among these, nitric oxide is considered as one of the effective ways to reduce the adverse effects of metal toxicity in plants (Cerana and Malerba, 2015; Yang et al., 2016; Terrón-Camero et al., 2019).

## CONCLUSION

Nitric oxide helps to reduce metal toxicity by strengthening the antioxidant defense system. Nitric oxide encourages the regulation of physiological induces and activation of plant defense system. The elimination of toxic ROS was recorded in the stressed plants when treated with nitric oxide. Photosynthetic pigments restore through nitric oxide supplementation. To overcome metal toxicity, plant growth and yield can be improved with exogenous nitric oxide supplementation. In future, development of tolerant genetic resources is need of time for higher productivity of crops.

## REFERENCES

Ahmad, A., Khan, W. U., Shah, A. A., Yasin, N. A., Naz, S., Ali, A., Tahir, A., & Batool, A. I. (2021). Synergistic effects of nitric oxide and silicon on promoting plant growth, oxidative stress tolerance and reduction of arsenic uptake in Brassica juncea. *Chemosphere*, *262*, 128384. DOI: 10.1016/j.chemosphere.2020.128384

Ahmad, P., Ahanger, M. A., Alyemeni, M. N., Wijaya, L., & Alam, P. (2018). Exogenous application of nitric oxide modulates osmolyte metabolism, antioxidants, enzymes of ascorbate-glutathione cycle and promotes growth under cadmium stress in tomato. *Protoplasma*, *255*(1), 79–93. DOI: 10.1007/s00709-017-1132-x

Ahmad, P., Alyemeni, M. N., Wijaya, L., Ahanger, M. A., Ashraf, M., Alam, P., Paray, B. A., & Rinklebe, J. (2021). Nitric oxide donor, sodium nitroprusside, mitigates mercury toxicity in different cultivars of soybean. *Journal of Hazardous Materials*, *408*, 124852. DOI: 10.1016/j.jhazmat.2020.124852

Ahmad, P., Alyemeni, M. N., Wijaya, L., Ahanger, M. A., Ashraf, M., Alam, P., Paray, B. A., & Rinklebe, J. (2021). Nitric oxide donor, sodium nitroprusside, mitigates mercury toxicity in different cultivars of soybean. *Journal of Hazardous Materials*, *408*, 124852. DOI: 10.1016/j.jhazmat.2020.124852

Altaf, M. A., Hao, Y., He, C., Mumtaz, M. A., Shu, H., Fu, H., & Wang, Z. (2022). Physiological and biochemical responses of pepper (Capsicum annuum L.) seedlings to nickel toxicity. *Frontiers in Plant Science*, *13*, 950392. DOI: 10.3389/fpls.2022.950392

Angulo-Bejarano, P. I., Puente-Rivera, J., & Cruz-Ortega, R. (2021). Metal and metalloid toxicity in plants: An overview on molecular aspects. *Plants*, *10*(4), 635. DOI: 10.3390/plants10040635

Arif, M. S., Yasmeen, T., Abbas, Z., Ali, S., Rizwan, M., Aljarba, N. H., Alkahtani, S., & Abdel-Daim, M. M. (2021). Role of exogenous and endogenous hydrogen sulfide (H2S) on functional traits of plants under heavy metal stresses: A recent perspective. *Frontiers in Plant Science*, *11*, 545453. DOI: 10.3389/fpls.2020.545453

Astier, J., Gross, I., & Durner, J. (2018). Nitric oxide production in plants: An update. *Journal of Experimental Botany*, *69*(14), 3401–3411. DOI: 10.1093/jxb/erx420

Basit, F., Abbas, S., Sheteiwy, M. S., Bhat, J. A., Alsahli, A. A., & Ahmad, P. (2024). Deciphering the alleviation potential of nitric oxide, for low temperature and chromium stress via maintaining photosynthetic capacity, antioxidant defence, and redox homeostasis in rice (Oryza sativa). *Plant Physiology and Biochemistry : PPB*, *214*, 108957. DOI: 10.1016/j.plaphy.2024.108957

Basit, F., Bhat, J. A., Guan, Y., Jan, B. L., Tyagi, A., & Ahmad, P. (2022). Nitric oxide and spermine revealed positive defense interplay for the regulation of the chromium toxicity in soybean (Glycine max L.). *Environmental Pollution*, *308*, 119602. DOI: 10.1016/j.envpol.2022.119602

Begara-Morales, J. C., Chaki, M., Valderrama, R., Sánchez-Calvo, B., Mata-Pérez, C., Padilla, M. N., Corpas, F. J., & Barroso, J. B. (2018). Nitric oxide buffering and conditional nitric oxide release in stress response. *Journal of Experimental Botany*, *69*(14), 3425–3438. DOI: 10.1093/jxb/ery072

Buet, A., Galatro, A., Ramos-Artuso, F., & Simontacchi, M. (2019). Nitric oxide and plant mineral nutrition: Current knowledge. *Journal of Experimental Botany*, *70*(17), 4461–4476. DOI: 10.1093/jxb/erz129

Buet, A., Moriconi, J. I., Santa-María, G. E., & Simontacchi, M. (2014). An exogenous source of nitric oxide modulates zinc nutritional status in wheat plants. *Plant Physiology and Biochemistry : PPB*, *83*, 337–345. DOI: 10.1016/j.plaphy.2014.08.020

Castello, D. F., Nejamkin, A., Cassia, R., Correa-Aragunde, N., Fernández, B., Foresi, N., Lombardo, C., Ramirez, L., & Lamattina, L. (2019). The era of nitric oxide in plant biology: Twenty years tying up loose ends. *Nitric Oxide : Biology and Chemistry*, *85*, 17–27. DOI: 10.1016/j.niox.2019.01.013

Cerana, R., & Malerba, M. (2015). Role of nitric oxide in heavy metal stress. In *Nitric oxide action in abiotic stress responses in plants* (pp. 181–192). Springer International Publishing. DOI: 10.1007/978-3-319-17804-2_12

Clemens, S. (2006). Toxic metal accumulation, responses to exposure and mechanisms of tolerance in plants. *Biochimie*, *88*(11), 1707–1719. DOI: 10.1016/j.biochi.2006.07.003

Dash, L., Singh, A., & Roychoudhury, A. (2025). Non-enzymatic antioxidant defense systems in plants. In *Role of Antioxidants in Mitigating Plant Stress* (pp. 51–61). Academic Press. DOI: 10.1016/B978-0-443-26799-4.00013-X

Duan, X., Li, X., Ding, F., Zhao, J., Guo, A., Zhang, L., Yao, J., & Yang, Y. (2015). Interaction of nitric oxide and reactive oxygen species and associated regulation of root growth in wheat seedlings under zinc stress. *Ecotoxicology and Environmental Safety*, *113*, 95–102. DOI: 10.1016/j.ecoenv.2014.11.030

Emamverdian, A., Ding, Y., Barker, J., Mokhberdoran, F., Ramakrishnan, M., Liu, G., & Li, Y. (2021). Nitric oxide ameliorates plant metal toxicity by increasing antioxidant capacity and reducing Pb and Cd translocation. *Antioxidants*, *10*(12), 1981. DOI: 10.3390/antiox10121981

Freschi, L. (2013). Nitric oxide and phytohormone interactions: Current status and perspectives. *Frontiers in Plant Science*, *4*, 398. DOI: 10.3389/fpls.2013.00398

Ghorbani, A., Pishkar, L., Roodbari, N., Pehlivan, N., & Wu, C. (2021). Nitric oxide could allay arsenic phytotoxicity in tomato (Solanum lycopersicum L.) by modulating photosynthetic pigments, phytochelatin metabolism, molecular redox status and arsenic sequestration. *Plant Physiology and Biochemistry : PPB*, *167*, 337–348. DOI: 10.1016/j.plaphy.2021.08.019

Ghuge, S. A., Nikalje, G. C., Kadam, U. S., Suprasanna, P., & Hong, J. C. (2023). Comprehensive mechanisms of heavy metal toxicity in plants, detoxification, and remediation. *Journal of Hazardous Materials, 450*, 131039. DOI: 10.1016/j.jhazmat.2023.131039

Gill, S. S., Hasanuzzaman, M., Nahar, K., Macovei, A., & Tuteja, N. (2013). Importance of nitric oxide in cadmium stress tolerance in crop plants. *Plant Physiology and Biochemistry : PPB*, *63*, 254–261. DOI: 10.1016/j.plaphy.2012.12.001

Graziano, M., & Lamattina, L. (2005). Nitric oxide and iron in plants: An emerging and converging story. *Trends in Plant Science*, *10*(1), 4–8. DOI: 10.1016/j.tplants.2004.12.004

Gupta, K. J., Kaladhar, V. C., Fitzpatrick, T. B., Fernie, A. R., Møller, I. M., & Loake, G. J. (2022). Nitric oxide regulation of plant metabolism. *Molecular Plant*, *15*(2), 228–242. DOI: 10.1016/j.molp.2021.12.012

Hashem, H. A., Hassanein, A. A., & Esmail, N. Y. (2018). Nitric oxide enhances the adaptive responses of lupine plants against heavy-metal stress. *Australian Journal of Crop Science*, *12*(12), 1962–1975. DOI: 10.21475/ajcs.18.12.12.p1379

Hoque, M. N., Tahjib-Ul-Arif, M., Hannan, A., Sultana, N., Akhter, S., Hasanuzzaman, M., Akter, F., Hossain, M. S., Sayed, M. A., Hasan, M. T., Skalicky, M., Li, X., & Brestič, M. (2021). Melatonin modulates plant tolerance to heavy metal stress: Morphological responses to molecular mechanisms. *International Journal of Molecular Sciences*, *22*(21), 11445. DOI: 10.3390/ijms222111445

Huihui, Z., Xin, L., Zisong, X., Yue, W., Zhiyuan, T., Meijun, A., Yuehui, Z., Wenxu, Z., Nan, X., & Guangyu, S. (2020). Toxic effects of heavy metals Pb and Cd on mulberry (Morus alba L.) seedling leaves: Photosynthetic function and reactive oxygen species (ROS) metabolism responses. *Ecotoxicology and Environmental Safety*, *195*, 110469. DOI: 10.1016/j.ecoenv.2020.110469

Imran, M., Hussain, S., Iqbal, A., Saleem, M. H., Rehman, N., Mo, Z., Chen, X., & Tang, X. (2023). Nitric oxide confers cadmium tolerance in fragrant rice by modulating physio-biochemical processes, yield attributes, and grain quality traits. *Ecotoxicology and Environmental Safety, 261*, 115078. DOI: 10.1016/j.ecoenv.2023.115078

Kaya, C., Akram, N. A., Sürücü, A., & Ashraf, M. (2019). Alleviating effect of nitric oxide on oxidative stress and antioxidant defence system in pepper (Capsicum annuum L.) plants exposed to cadmium and lead toxicity applied separately or in combination. *Scientia Horticulturae, 255*, 52–60. DOI: 10.1016/j.scienta.2019.05.029

Kaya, C., Ugurlar, F., Ashraf, M., El-Sheikh, M. A., Bajguz, A., & Ahmad, P. (2022). The participation of nitric oxide in hydrogen sulphide-mediated chromium tolerance in pepper (Capsicum annuum L) plants by modulating subcellular distribution of chromium and the ascorbate-glutathione cycle. *Environmental Pollution, 313*, 120229. DOI: 10.1016/j.envpol.2022.120229

Khan, E.A., Ahmed, H.M., Misra, M., Sharma, P. and Misra, A.N., 2022. Nitric Oxide Alleviates Photochemical Damage Induced by Cadmium Stress in Pea Seedlings. *Phyton (0031-9457), 91*(5).

Körpe, D. A., & Aras, S. (2011). Evaluation of copper-induced stress on eggplant (Solanum melongena L.) seedlings at the molecular and population levels by use of various biomarkers. *Mutation Research. Genetic Toxicology and Environmental Mutagenesis, 719*(1-2), 29–34. DOI: 10.1016/j.mrgentox.2010.10.003

Krasylenko, Y. A., Yemets, A. I., & Blume, Y. B. (2010). Functional role of nitric oxide in plants. *Russian Journal of Plant Physiology: a Comprehensive Russian Journal on Modern Phytophysiology, 57*(4), 451–461. DOI: 10.1134/S1021443710040011

Kumar, S., Shah, S. H., Vimala, Y., Jatav, H. S., Ahmad, P., Chen, Y., & Siddique, K. H. (2022). Abscisic acid: Metabolism, transport, crosstalk with other plant growth regulators, and its role in heavy metal stress mitigation. *Frontiers in Plant Science, 13*, 972856. DOI: 10.3389/fpls.2022.972856

Küpper, H., & Andresen, E. (2016). Mechanisms of metal toxicity in plants. *Metallomics : Integrated Biometal Science*, *8*(3), 269–285. DOI: 10.1039/C5MT00244C

Lee, M. S., Methela, N. J., Lee, G. H., & Mun, B. G. (2025). Nitric Oxide and Melatonin Cross Talk on Photosynthetic Machinery. *Molecules (Basel, Switzerland)*, *30*(10), 2148. DOI: 10.3390/molecules30102148

Liu, X., Gong, D., Ke, Q., Yin, L., Wang, S., & Gao, T. (2023). Meta-analysis of the effect of nitric oxide application on heavy metal stress tolerance in plants. *Plants*, *12*(7), 1494. DOI: 10.3390/plants12071494

Liu, X., Yin, L., Deng, X., Gong, D., Du, S., Wang, S., & Zhang, Z. (2020). Combined application of silicon and nitric oxide jointly alleviated cadmium accumulation and toxicity in maize. *Journal of Hazardous Materials*, *395*, 122679. DOI: 10.1016/j.jhazmat.2020.122679

Liu, X., Yin, L., Deng, X., Gong, D., Du, S., Wang, S., & Zhang, Z. (2020). Combined application of silicon and nitric oxide jointly alleviated cadmium accumulation and toxicity in maize. *Journal of Hazardous Materials*, *395*, 122679. DOI: 10.1016/j.jhazmat.2020.122679

Liu, Y., & Zhang, H. (2021). Reactive oxygen species and nitric oxide as mediators in plant hypersensitive response and stomatal closure. *Plant Signaling & Behavior*, *16*(12), 1985860. DOI: 10.1080/15592324.2021.1985860

Mahjoubi, Y., Emmerson, R., Kharbech, O., Massoud, M. B., Davey, P., & Djebali, W. (2025). Nitric oxide application improves photosynthetic light-response through enhancing photosystem II efficiency of bean plants (Phaseolus vulgaris L.) submitted to manganese stress. *Acta Physiologiae Plantarum*, *47*(2), 24. DOI: 10.1007/s11738-025-03779-4

Mandal, A., Mundhe, D., Sonkamble, V., Wagh, N., & Lakkakula, J. (2022). Impact of heavy metal contamination on soil environment and advances in its revitalization strategies. In *Development in Waste Water Treatment Research and Processes* (pp. 215–241). Elsevier. DOI: 10.1016/B978-0-323-85584-6.00014-5

Mata-Pérez, C., Sánchez-Vicente, I., Arteaga, N., Gómez-Jiménez, S., Fuentes-Terrón, A., Oulebsir, C. S., Calvo-Polanco, M., Oliver, C., & Lorenzo, Ó. (2023). Functions of nitric oxide-mediated post-translational modifications under abiotic stress. *Frontiers in Plant Science*, *14*, 1158184. DOI: 10.3389/fpls.2023.1158184

Melebary, S. J. (2023). Heavy metal toxicity and remediation in human and agricultural systems: A review. *Advances in Animal and Veterinary Sciences*, *11*(4), 679–694. DOI: 10.17582/journal.aavs/2023/11.4.679.694

Meng, Y., Jing, H., Huang, J., Shen, R., & Zhu, X. (2022). The role of nitric oxide signaling in plant responses to cadmium stress. *International Journal of Molecular Sciences*, *23*(13), 6901. DOI: 10.3390/ijms23136901

Mishra, V., Singh, P., Tripathi, D. K., Corpas, F. J., & Singh, V. P. (2021). Nitric oxide and hydrogen sulfide: An indispensable combination for plant functioning. *Trends in Plant Science*, *26*(12), 1270–1285. DOI: 10.1016/j.tplants.2021.07.016

Mushtaq, R., Shafiq, M., Batool, A., Din, M. I., & Sami, A. (2024). Investigate the impact of zinc oxide nanoparticles under lead toxicity on Chilli (Capsicum annuum L). *Bulletin of Biological and Allied Sciences Research*, *2024*(1), 90–90. DOI: 10.54112/bbasr.v2024i1.90

Nahar, K., Rhaman, M. S., Parvin, K., Bardhan, K., Marques, D. N., García-Caparrós, P., & Hasanuzzaman, M. (2022). Arsenic-induced oxidative stress and antioxidant defense in plants. *Stresses*, *2*(2), 179–209. DOI: 10.3390/stresses2020013

Nawaz, M., Saleem, M. H., Khalid, M. R., Ali, B., & Fahad, S. (2024). Nitric oxide reduces cadmium uptake in wheat (Triticum aestivum L.) by modulating growth, mineral uptake, yield attributes, and antioxidant profile. *Environmental Science and Pollution Research International*, *31*(6), 9844–9856. DOI: 10.1007/s11356-024-31875-4

Okant, M., & Kaya, C. (2019). The role of endogenous nitric oxide in melatonin-improved tolerance to lead toxicity in maize plants. *Environmental Science and Pollution Research International*, *26*(12), 11864–11874. DOI: 10.1007/s11356-019-04517-3

Okant, M., & Kaya, C. (2019). The role of endogenous nitric oxide in melatonin-improved tolerance to lead toxicity in maize plants. *Environmental Science and Pollution Research International, 26*(12), 11864–11874. DOI: 10.1007/s11356-019-04517-3

Qin, C., Shen, J., & Ahanger, M. A. (2022). Supplementation of nitric oxide and spermidine alleviates the nickel stress-induced damage to growth, chlorophyll metabolism, and photosynthesis by upregulating ascorbate–glutathione and glyoxalase cycle functioning in tomato. *Frontiers in Plant Science, 13*, 1039480. DOI: 10.3389/fpls.2022.1039480

Ran, S., Zhang, K., Zhou, Y., Huang, W., & Zhong, F. (2023). Nickel Ions Enhanced the Adaptability of Tomato Seedling Roots to Low-Nitrogen Stress by Improving Their Antioxidant Capacity. *Horticulturae, 9*(12), 1342. DOI: 10.3390/horticulturae9121342

Rashid, A., Schutte, B. J., Ulery, A., Deyholos, M. K., Sanogo, S., Lehnhoff, E. A., & Beck, L. (2023). Heavy metal contamination in agricultural soil: Environmental pollutants affecting crop health. *Agronomy (Basel), 13*(6), 1521. DOI: 10.3390/agronomy13061521

Raza, A., Hussain, S., Javed, R., Hafeez, M. B., & Hasanuzzaman, M. (2021). Antioxidant defense systems and remediation of metal toxicity in plants. In *Approaches to the remediation of inorganic pollutants* (pp. 91–124). Springer Singapore. DOI: 10.1007/978-981-15-6221-1_6

Raza, A., Salehi, H., Rahman, M. A., Zahid, Z., Madadkar Haghjou, M., Najafi-Kakavand, S., Charagh, S., Osman, H. S., Albaqami, M., Zhuang, Y., Siddique, K. H., & Zhuang, W. (2022). Plant hormones and neurotransmitter interactions mediate antioxidant defenses under induced oxidative stress in plants. *Frontiers in Plant Science, 13*, 961872. DOI: 10.3389/fpls.2022.961872

Riyazuddin, R., Nisha, N., Ejaz, B., Khan, M. I. R., Kumar, M., Ramteke, P. W., & Gupta, R. (2021). A comprehensive review on the heavy metal toxicity and sequestration in plants. *Biomolecules, 12*(1), 43. DOI: 10.3390/biom12010043

Sami, F., Faizan, M., Faraz, A., Siddiqui, H., Yusuf, M., & Hayat, S. (2018). Nitric oxide-mediated integrative alterations in plant metabolism to confer abiotic stress tolerance, NO crosstalk with phytohormones and NO-mediated post translational modifications in modulating diverse plant stress. *Nitric Oxide : Biology and Chemistry*, *73*, 22–38. DOI: 10.1016/j.niox.2017.12.005

Saxena, I., & Shekhawat, G. S. (2013). Nitric oxide (NO) in alleviation of heavy metal induced phytotoxicity and its role in protein nitration. *Nitric Oxide : Biology and Chemistry*, *32*, 13–20. DOI: 10.1016/j.niox.2013.03.004

Saxena, I., & Shekhawat, G. S. (2013). Nitric oxide (NO) in alleviation of heavy metal induced phytotoxicity and its role in protein nitration. *Nitric Oxide : Biology and Chemistry*, *32*, 13–20. DOI: 10.1016/j.niox.2013.03.004

Seifried, H. E., Anderson, D. E., Fisher, E. I., & Milner, J. A. (2007). A review of the interaction among dietary antioxidants and reactive oxygen species. *The Journal of Nutritional Biochemistry*, *18*(9), 567–579. DOI: 10.1016/j.jnutbio.2006.10.007

Sharma, A., Soares, C., Sousa, B., Martins, M., Kumar, V., Shahzad, B., Sidhu, G. P., Bali, A. S., Asgher, M., Bhardwaj, R., Thukral, A. K., Fidalgo, F., & Zheng, B. (2020). Nitric oxide-mediated regulation of oxidative stress in plants under metal stress: A review on molecular and biochemical aspects. *Physiologia Plantarum*, *168*(2), 318–344. DOI: 10.1111/ppl.13004

Siddiqui, M. H., Al-Whaibi, M. H., Ali, H. M., Sakran, A. M., Basalah, M. O., & AlKhaishany, M. Y. (2013). Mitigation of nickel stress by the exogenous application of salicylic acid and nitric oxide in wheat. *Australian Journal of Crop Science*, *7*(11), 1780–1788.

Singh, H. P., Kaur, S., Batish, D. R., Sharma, V. P., Sharma, N., & Kohli, R. K. (2009). Nitric oxide alleviates arsenic toxicity by reducing oxidative damage in the roots of Oryza sativa (rice). *Nitric Oxide : Biology and Chemistry*, *20*(4), 289–297. DOI: 10.1016/j.niox.2009.02.004

Singhal, R. K., Jatav, H. S., Aftab, T., Pandey, S., Mishra, U. N., Chauhan, J., Chand, S., Indu, , Saha, D., Dadarwal, B. K., Chandra, K., Khan, M. A., Rajput, V. D., Minkina, T., Narayana, E. S., Sharma, M. K., & Ahmed, S. (2021). Roles of nitric oxide in conferring multiple abiotic stress tolerance in plants and crosstalk with other plant growth regulators. *Journal of Plant Growth Regulation*, *40*(6), 2303–2328. DOI: 10.1007/s00344-021-10446-8

Širić, I., Eid, E. M., El-Morsy, M. H., Osman, H. E., Adelodun, B., Abou Fayssal, S., Mioč, B., Goala, M., Singh, J., Bachheti, A., Arya, A. K., Choi, K. S., Kumar, V., & Kumar, P. (2022). Health risk assessment of hazardous heavy metals in two varieties of mango fruit (Mangifera indica L. var. Dasheri and Langra). *Horticulturae*, *8*(9), 832. DOI: 10.3390/horticulturae8090832

Soliman, M., Alhaithloul, H. A., Hakeem, K. R., Alharbi, B. M., El-Esawi, M., & Elkelish, A. (2019). Exogenous nitric oxide mitigates nickel-induced oxidative damage in eggplant by upregulating antioxidants, osmolyte metabolism, and glyoxalase systems. *Plants*, *8*(12), 562. DOI: 10.3390/plants8120562

Tak, H. I., Ahmad, F., & Babalola, O. O. (2012). Advances in the application of plant growth-promoting rhizobacteria in phytoremediation of heavy metals. *Reviews of Environmental Contamination and Toxicology*, *223*, 33–52.

Terrón-Camero, L. C., Peláez-Vico, M. Á., Del-Val, C., Sandalio, L. M., & Romero-Puertas, M. C. (2019). Role of nitric oxide in plant responses to heavy metal stress: Exogenous application versus endogenous production. *Journal of Experimental Botany*, *70*(17), 4477–4488. DOI: 10.1093/jxb/erz184

Verma, N., & Prasad, S. M. (2021). Interplay of hydrogen peroxide and nitric oxide: Systemic regulation of photosynthetic performance and nitrogen metabolism in cadmium challenged cyanobacteria. *Physiology and Molecular Biology of Plants : an International Journal of Functional Plant Biology*, *27*(10), 2181–2199. DOI: 10.1007/s12298-021-01083-2

Verma, N., Tiwari, S., Singh, V. P., & Prasad, S. M. (2020). Nitric oxide in plants: An ancient molecule with new tasks. *Plant Growth Regulation*, *90*(1), 1–13. DOI: 10.1007/s10725-019-00543-w

Wang, R., Huang, J., Liang, A., Wang, Y., Mur, L. A. J., Wang, M., & Guo, S. (2020). Zinc and copper enhance cucumber tolerance to fusaric acid by mediating its distribution and toxicity and modifying the antioxidant system. *International Journal of Molecular Sciences*, *21*(9), 3370. DOI: 10.3390/ijms21093370

Wei, L., Zhang, J., Wang, C., & Liao, W. (2020). Recent progress in the knowledge on the alleviating effect of nitric oxide on heavy metal stress in plants. *Plant Physiology and Biochemistry : PPB*, *147*, 161–171. DOI: 10.1016/j.plaphy.2019.12.021

Xiong, J., Fu, G., Tao, L., & Zhu, C. (2010). Roles of nitric oxide in alleviating heavy metal toxicity in plants. *Archives of Biochemistry and Biophysics*, *497*(1-2), 13–20. DOI: 10.1016/j.abb.2010.02.014

Yang, L., Ji, J., Harris-Shultz, K. R., Wang, H., Wang, H., Abd-Allah, E. F., Luo, Y., & Hu, X. (2016). The dynamic changes of the plasma membrane proteins and the protective roles of nitric oxide in rice subjected to heavy metal cadmium stress. *Frontiers in Plant Science*, *7*, 190. DOI: 10.3389/fpls.2016.00190

Zhang, J., Zeng, L., Sun, H., Wu, H., & Chen, S. (2017). Adversity stress-related responses at physiological attributes, transcriptional and enzymatic levels after exposure to Cu in Lycopersicum esculentm seedlings. *Scientia Horticulturae*, *222*, 213–220. DOI: 10.1016/j.scienta.2017.05.027

Zhang, Y., Wang, R., Wang, X., Zhao, C., Shen, H., & Yang, L. (2023). Nitric oxide regulates seed germination by integrating multiple signalling pathways. *International Journal of Molecular Sciences*, *24*(10), 9052. DOI: 10.3390/ijms24109052

Zhang, Y., Yang, C., Liu, S., Xie, Z., Chang, H., & Wu, T. (2024). Phytohormones-mediated strategies for mitigation of heavy metals toxicity in plants focused on sustainable production. *Plant Cell Reports*, *43*(4), 99. DOI: 10.1007/s00299-024-03189-9

# Chapter 5
# Plant Response to Jasmonates Under Metal Toxicity by Alteration of Morpho-Physiological and Biochemical Insights

**Zarina Bibi**
http://orcid.org/0000-0002-9016-1697
*The University of Agriculture, Dera Ismail Khan, Pakistan*

**Mohamed A. A. Ahmed**
*Alexandria University, Egypt*

## ABSTRACT

*Heavy metal deposition in soils can negatively impact crop growth and development owing to phytotoxicity, food safety and marketability, and the environmental health of soil organisms. It has become an issue in agricultural production. Plants' metabolic processes have an impact on the biological and geological dispersion of heavy metals by contaminating the air, water, and soil. Plants are an essential part of the ecosystem. Metal buildup, reduced development, altered metabolism, and decreased*

DOI: 10.4018/979-8-3693-9899-9.ch005

*biomass production are all observed in plants living in metal-polluted environments. Metals have an impact on many physiological and biochemical processes in plants. Plant responses to metal toxicity and plant growth and development are regulated by jasmonates (JAs), which are widely distributed lipid-derived chemicals. Jasmonic acid belongs to the oxylipin family, as are its several metabolites. Hence, it has been illustrated that jasmonic acid plays a significant role in improving crops' tolerance to physiological and biochemical mechanisms that lessens the adverse effects of metal toxicity.*

## INTRODUCTION

Metal toxicity is a serious threat to the sustainability of ecosystems, agricultural output, and plant health. The optimum levels of metals are necessary for sustainable plant growth and yield. The higher concentrations of metals are toxic to plant health (Zhang et al., 2024). The higher concentrations may possibly be due to human activities, industrial discharge, excessive use of agrochemicals, and the utilization of sewage sludge (Saini et al., 2021). These hazardous metals interfere with plant metabolic activities, causing oxidative harm, changing nutrient intake, destroying cellular structures, and disrupting physiological processes (Riyazuddin et al., 2021). Phytohormones reduce the harmful effects of heavy metals (Bücker-Neto et al., 2017). Jasmonic acid (JA) and its derivatives, referred to as jasmonates (JAs), have illustrated a significant impact in improving plant tolerance against metal toxicity. Growth inhibition, decreased biomass output, chlorosis of the leaves, nutrient and water imbalance, and plant senescence are only a few of the detrimental impacts that heavy metals have on plants (Mousavi et al., 2020). However, despite their detrimental effects on plants, metals are still considered a serious health risk to humans due to their persistence in the environment (Bücker-Neto et al., 2017). Metals have a significant impact on crop plants, which ultimately reduces the total productivity of the plant. Furthermore, plants growing in contaminated soils accumulate a lot of metals, which causes them to enter the food chain and result in infections (Hasan et al., 2021; Angon et al., 2024). The multifunctional metal transport systems seen in plants move and gather the metals within their cells. To increase the conversion of metallic ions, the appearance of the plant protein family's ATP-binding cassette (ABS), which is involved in ion movement, might be changed (Sharma and Agrawal, 2005). The formation of various proteins and secondary metabolites, changes in hormone signaling, gene expression, and antioxidant defense activities can all be disrupted by the severity of metals (Rai et al., 2019). Among other physiological advancements, antioxidant enzymes and phytohormones are crucial in determining the molecular expression of the plant's genes (Jaiswal et al., 2018). Phytohormones are signaling molecules that control various plant reactions in adverse conditions (Rashid et al., 2023). JAs are considered among the most well-studied phytohormones. Previously identified as stress-related hormones in higher plants, JAs are endogenous signaling molecules involved in several metabolic processes (Khan et al., 2015). JA is developed for evolutionary goals in plants and initiates defense responses against metal stress (Zhou et al., 2013). Meanwhile, exogenous JA and methyl jasmonate (MeJA) enhance antioxidant defense against metal toxicity (Rehman et al., 2023; El-Mogy et al., 2025). Jasmonates improves metal tolerance in maize (Mir et al., 2018), wheat (Ahmad et al., 2021), rice (Li et al., 2024), tomato (Pascual et al., 2023), and cucumber (Wang et al., 2022). Present chapter explore the beneficial role that JA plays under metal toxicity to improve metal tolerance in plants. Furthermore, the antioxidant defense system through JA and its interaction with other phytohormones were also elucidated in plants.

## CONSEQUENCES OF JAS IN PLANT GROWTH AND PRODUCTIVITY

JAs are essential for plant growth, development, and stress responses (Wasternack and Hause, 2013). Several developmental processes, such as petal growth, root elongation, and seed germination, were regulated with JAs (Sohn et al., 2022). JAs signaling affects gene expression and metabolic processes through intricate networks and associations with phytohormones (Wasternack and Hause, 2013). Important elements of JAs sensing and signaling, including proteins and their interactors, have been explored by several plant researchers (Liu and Timko, 2021). JAs are engaged in growth-defense compromises, in which induced tolerance frequently results in slower development. This balance might maximize the expression of defensive chemicals but without losing vital metabolic processes (Hewedy et al., 2023). Furthermore, JAs function as efficient elicitors for the production of secondary metabolites (Nguyen et al., 2022). Plant developmental activities like seed germination, root growth, leaf senescence, floral development, and fruit ripening are all greatly impacted by JAs. However, interactions with other phytohormones might affect their effects, which are frequently concentration-dependent (Huang et al., 2017).

JAs may improve the uptake of water and nutrients by encouraging the growth of adventitious roots and the formation of root hairs even at low concentrations. However, higher levels of JAs can prevent primary root extension, mostly by preventing cell proliferation in the root meristem (Shikha et al., 2023). Crosstalk of JAs with auxin transport and signaling pathways is frequently initiated. JAs affect the photosynthetic capacity and longevity of leaves by accelerating the last stage of leaf growth, known as senescence, through the control of senescence-associated genes (Jang et al., 2020). JAs have a crucial role in flower opening, anther dehiscence, and pollen maturation during reproductive development. Their crucial involvement in fertility and effective reproduction is highlighted by the male sterility shown in JA-deficient mutants, such as the Arabidopsis *opr3* and *dde2* mutants (Avanci et al., 2010). Jasmonic acid (JA) and methyl jasmonate (MeJA) control the growth, development, and defense responses of plants (Liu and Timko, 2021). More and more data in recent years have shown the significance of JAs for improving plant tolerance to metal toxicity (Manzoor et al., 2025). Because of their ability to cause oxidative stress, interfere with metabolic processes, and prevent photosynthesis and nutrient uptake, metal toxicity represents a serious risk to plant productivity (Naz et al., 2022; Muhammad et al., 2025). JAs serve as essential modulators in the complex signaling networks that enable plants to recognize, respond, and adapt to these harmful metal circumstances (He et al., 2002).

The overproduction of free radicals and toxic substances occurs in peppers under metal stress and can lead to protein breakdown, DNA damage, and lipid peroxidation. JAs lessen the adverse effects of metal toxicity by strengthening the antioxidant defense systems of the plant (Zhang et al., 22023). This enzymatic antioxidant system helps in scavenging toxic elements and defending cellular structures under stress environments. Moreover, JAs help with metal sequestration and detoxification by controlling the expression of chelating molecules, including phytochelatins and metallothionein, as well as metal transporter genes. For instance, exogenous application of MeJA reduces the development of Cd and Pb in plant tissues by preventing their transfer from roots to shoots, thereby maintaining photosynthetic organs of peppers (Ekinci et al., 2024). Additionally, it can boost the production of non-enzymatic antioxidants that are essential for preserving redox equilibrium, such as glutathione (GSH) and ascorbic acid (AsA) (Ahmad et al., 2024; Muhammad et al., 2024). Heavy metals harm the photosynthetic machinery and limit the manufacture of chlorophyll. Nonetheless, under stress, plants treated with JAs frequently exhibit increased net photosynthetic rates, better stomatal control, and higher chlorophyll content (Altaf et al., 2023; El-Saadony et al., 2024). When compared to untreated plants, this preservation of photosynthesis results in improved biomass accumulation and yield maintenance. JAs treatment enhances root architecture under metal toxicity by modifying root elongation, lateral root creation, and root hair development (Altaf et al., 2024). Even in soils polluted with metals, this adaptive response promotes improved uptake of water and nutrients. Furthermore, genes involved in the synthesis of secondary metabolites, such as flavonoids and phenolics, which have antioxidant and metal-chelating qualities, are expressed more frequently when JA signaling is present (Hu et al., 2020; Li et al., 2021; Abdollahi et al., 2023; Song et al., 2026). These substances help the plant tolerate stress in general while also lowering the bioavailability of metals. Jasmonates enhance plant growth and yield by regulating the photosynthetic system (Figure 1).

*Figure 1. Jasmonates enhance plant growth and yield by regulation of photosynthetic system under metal toxicity*

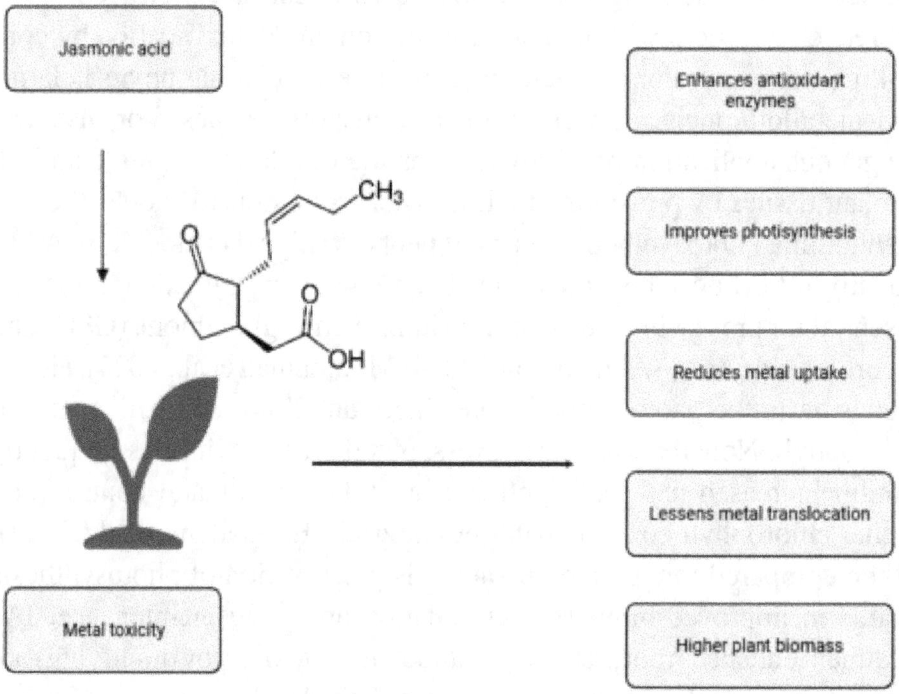


## JAS REGULATE THE PHOTOSYNTHETIC SYSTEM IN PLANTS UNDER METAL TOXICITY

JAs are essential for controlling photosynthetic systems and increasing plant resistance to heavy metal toxicity. Under metal stress, jasmonates preserve cell integrity, boost osmoprotectant production, and activate antioxidant defense systems (Huihui et al., 2020). They control the build-up of hazardous elements in plants by coordinating chelating capability and ion transport systems (Khan et al., 2019). JAs have varying impacts on light-harvesting antennae during metal stress in terms of chlorophyll accumulation, photosynthetic activity, and energy dissipation, depending on concentration and metal type (Niu et al., 2025). Exogenous use of JAs can increase plant tolerance to metal toxicity by regulating antioxidant metabolism, osmolyte production, and metabolite accumulation (Nazir et al., 2023). The significance of JAs in plant adaptation to environmental changes is highlighted by their interactions with other phytohormones as well as calcium signaling pathways to control plant responses to metal toxicity (Zhang et al., 2024).

*Figure 2. Jasmonates regulate photosynthesis in metal-stressed plants*

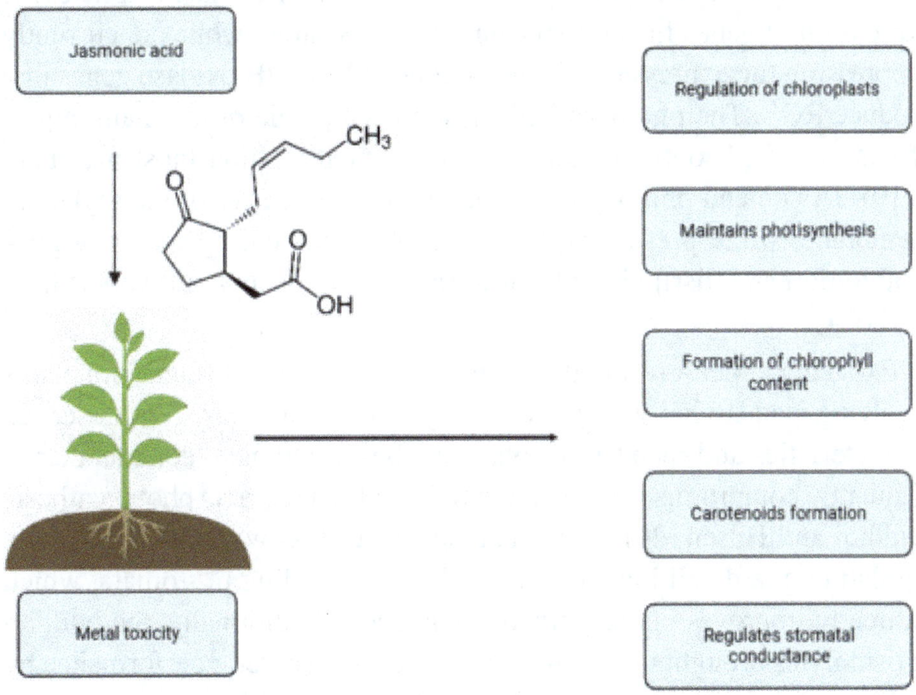


The overproduction of toxic substances and free radicals is one of the main causes of this impairment. ROS harm pigments, membranes, photosystems, and chloroplasts. Plants respond by activating various signaling pathways, and JAs are essential for reducing damage brought on by excess metals (Wang et al., 2020). JAs have become crucial modulators that control the manner in which plants react to abiotic stressors, particularly by preserving photosynthesis and controlling ROS homeostasis (Figure 2). ROS levels rise above the antioxidant system's capacity to detoxify metal toxicity. However, metals result in oxidative stress that impairs stomatal conductance, photosystem II, chlorophyll production, and carbon absorption (Ali and Baek, 2020). JAs lessen these adverse effects and lower the buildup of ROS by modifying both enzymatic and non-enzymatic antioxidant systems. To detoxify ROS and stop lipid peroxidation along with protein oxidation in chloroplast membranes, they trigger the activation of the plant defense system (Yan et al., 2015). JAs signaling modifies the redox balance and develops a complex signaling network that improves metal tolerance through interacting with other phytohormones. Shi et al. (2024) confirmed that applying JAs to heavy metal exposure increases the expression of genes linked to the antioxidant system, while concurrently suppressing the expression of stress-induced NADPH oxidase genes that produce ROS. The photosynthetic system is a single of the main targets of JAs-mediated control under aluminum toxicity (Ulloa-Inostroza et al., 2019). JAs defend photosystems, maintain the integrity of the thylakoid membrane, while preserving the chlorophyll content by reducing photoinhibition and disrupting electron transport in peanut cultivars (Shi et al., 2024).

Exogenous MeJA resulted in better photosynthetic efficiency in plants involving metal toxicity, as evidenced by raising chlorophyll fluorescence characteristics and recovery of photosynthesis. Stomatal conductance is frequently compromised by metal toxicity, which restricts photosynthesis (Siddiqi and Husen, 2019). JAs balance water loss with $CO_2$ uptake via regulating guard cell ion channels and ROS signaling in stomata, which indirectly improves photosynthetic efficiency. Additionally, the buildup of osmolytes brought on by JAs supports photosynthetic performance by aiding in osmotic adjustment and membrane stabilization during metal toxicity in pigeon pea plants (Kaushik et al., 2024). JAs preserve photosynthetic efficiency in several ways under metal toxicity. They ensure the sustainability of photosynthetic function by lowering ROS accumulation, boosting antioxidant defense, maintaining chloroplast structure, and controlling gene expression (Sharma et al., 2019). Their function as key regulators of plant defense is further supported by their interactions with other hormonal pathways. Agricultural production is still threatened by metal contamination; comprehending and modifying JA pathways presents viable methods for developing cultivars that can tolerate toxicity (Yu et al., 2018). Jasmonates regulate the photosynthetic mechanism in plants under metal toxicity (Table 1).

*Table 1. Jasmonates regulate the photosynthetic mechanism in plants under metal toxicity*

| Crop name | Metal type | Jasmonates impact | References |
|---|---|---|---|
| Wheat | Cadmium | Restores the membrane integrity | Alikhani and Abbaspour (2019) |
| Mustard | Arsenic | Maintains the photosynthesis system | Farooq et al. (2016) |
| Rice | Arsenic | Restores photochemical effectiveness | Verma et al. (2020) |
| Tomato | Nickel | Defends photosystems from oxidative injury | Yu et al. (2024) |
| Tomato | Cadmium | Maintains stomatal conductance | Wei et al. (2021) |
| Chili | Zinc | Boosts antioxidant potential in chloroplasts | Ahmed et al. (2024) |

## JAS COULD LESSEN THE TOXIC SUBSTANCES GENERATED IN PLANTS UNDER METAL TOXICITY

ROS are produced in excess by plants exposed to heavy metals, which causes oxidative stress, membrane lipid peroxidation, protein deterioration, and DNA damage. JAs, a type of phytohormone produced from lipids, modulate the antioxidant defense system to eliminate these toxic substances (Raza et al., 2021). To eliminate ROS, they increase the activity of important antioxidant enzymes. JAs also affect the expression of genes linked to detoxification pathways, such as the activation of metallothionein and glutathione S-transferases, that bound and sequester metals, thereby lowering their toxicity. Furthermore, JAs assist the production of non-enzymatic antioxidants such as ascorbate and glutathione, which support cellular redox balance (Sood, 2023). JAs improve plant health, longevity, and photosynthetic efficiency by drastically reducing the buildup and detrimental effects of toxic compounds produced during metal stress. JAs are essential for increasing a plant's resistance to the harmful effects of various metals. To lessen the negative effects of metals, they control several physiological, biochemical, and molecular processes (Raza et al., 2021). JAs enhance osmoprotectant production, boost JA-associated gene expression, and strengthen antioxidant defense mechanisms (Chen et al., 2021). Additionally, they improve plants' chelating capability and synchronize ion transport systems. Research has demonstrated that in plants under cadmium stress, MeJA increases the activities of antioxidant enzymes, while decreasing the levels of hydrogen peroxide and malondialdehyde (Keramat et al., 2009). JAs help reduce the toxicity of metals. The evolutionary conservation of JA-responsive gene families across major green plant lineages implies that JAs serve as essential phytohormones that shield plants from the harmful effects of metals and metalloids (Lei et al., 2020). Jasmonates reduce toxic elements produced in plants under metal toxicity (Table 2).

*Table 2. Jasmonates reduce toxic elements produced in plants under metal toxicity*

| Toxic elements | Jasmonates action | References |
|---|---|---|
| Reactive oxygen species (ROS) | Scavenging of free radicals and toxic elements | Zhang et al. (2023) |
| Hydrogen peroxide ($H_2O_2$) | Protects lipid peroxidation and membrane injury | Mehdizadeh et al. (2024) |
| Superoxide radical ($O_2^-$) | Shields chloroplasts and mitochondria from overgeneration of ROS injury | Shikha et al. (2023) |
| Malondialdehyde (MDA) | Depresses cell impairment and leakage | Yu et al. (2018) |
| Metal ions | Encourages phytochelatin synthesis as well as vacuolar sequestration | Mousavi et al. (2020) |

## JAS BOOST PLANT DEFENSE ACTIVITIES AGAINST METAL TOXICITY

JAs are essential for increasing a plant's resistance to the harmful effects of heavy metals. Being exposed to heavy metals, these phytohormones produced from lipids control plant growth and development (Yu et al., 2018). JAs strengthen plant defense mechanisms by increasing antioxidant systems, osmolyte production, and metabolite accumulation (Cheong and

Choi, 2003). Jas had the potential to lessen toxicity in combination with other phytohormones. JA treatment has been demonstrated to lessen lead toxicity among tomato seedlings by improving secondary metabolites, organic acids, and metal chelating chemicals (Mohamed and Latif, 2017). Furthermore, JAs can reduce the buildup of heavy metals by coordinating ion transport systems and boosting chelating ability in plants (Farooq et al., 2016; Per et al., 2016). JAs are essential for strengthening plants' defenses against metal toxicity. They operate as important signaling molecules that trigger enzymatic and non-enzymatic antioxidant mechanisms, reducing oxidative stress brought on by excess metals (Repkina et al., 2023). Important antioxidant enzymes that detoxify ROS and preserve cellular redox balance are produced and activated with JAs supplementation (Tu et al., 2025). JAs restrict the accumulation and mobility of metals in plant tissues by controlling the expression of genes involved in metal chelation, transport, and sequestration. Furthermore, they contribute to growth, membrane stability, and photosynthetic efficiency by lowering lipid peroxidation and safeguarding chloroplast structures (Zhang et al., 2024). JAs optimize the defense response and guarantee the best possible stress adaptation by interacting with other phytohormones (Wu et al., 2022). JAs greatly increase the plant's tolerance and resilience to metal toxicity by combining physiological, biochemical, and molecular defense mechanisms (Attia and Alamer, 2024). Jasmonates boost the antioxidant defense system of plants subjected to metal toxicity (Figure 3).

*Figure 3. Jasmonates boost antioxidant defense system of plants subjected to metal toxicity*

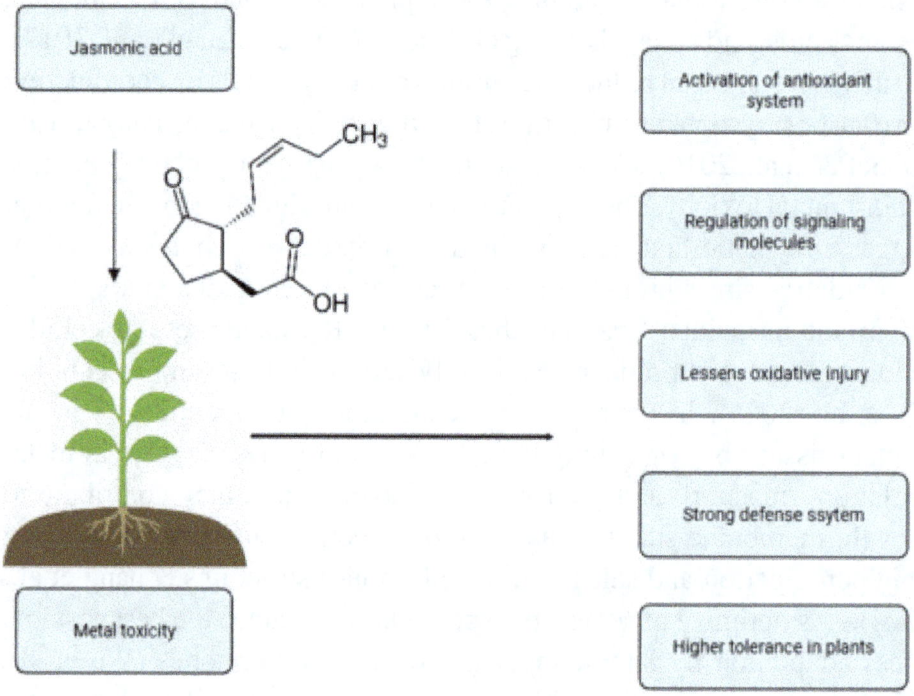


## CONCLUSION

Jasmonates (JAs) and their methyl esters have a significant impact on plant growth and development under metal toxicity. JAs, as growth regulators, have been shown to mitigate the negative consequences of metal toxicity by controlling various physiological and biochemical processes. JAs control the expression of genes involved in osmolyte biosynthesis, metabolite formation, antioxidant defense, general plant development, and physiological traits. Nevertheless, the analysis of plant HM signal detection, JAs biosynthesis, and response to several metals has not yet been fully recognized. JAs and other phytohormone signaling pathways may interact to control when plants respond to metal toxicity. Therefore, future studies that manipulate fundamental knowledge about JA's function and activation under metals may produce favorable outcomes. Further focus is also needed in the future on the JA signaling machinery and its functions in signaling cross-talk at the cellular level. Similarly, further research is still needed to understand the interaction of JAs and other phytohormones under metal toxicity.

## REFERENCES


Abdollahi, A., Farsad-Akhtar, N., Mohajel Kazemi, E., & Kolahi, M. (2023). Investigation of the combined effects of cadmium chloride, silver nitrate, lead nitrate, methyl jasmonate, and salicylic acid on morphometric and biochemical characteristics of St. John's wort. *Physiology and Molecular Biology of Plants : an International Journal of Functional Plant Biology*, *29*(2), 173–184. DOI: 10.1007/s12298-023-01281-0

Ahmad, P., Raja, V., Ashraf, M., Wijaya, L., Bajguz, A., & Alyemeni, M. N. (2021). Jasmonic acid (JA) and gibberellic acid (GA3) mitigated Cd-toxicity in chickpea plants through restricted cd uptake and oxidative stress management. *Scientific Reports*, *11*(1), 19768. DOI: 10.1038/s41598-021-98753-8

Ahmad, R., Muhammad, H. M. D., Naz, S., Tiwari, R. K., Lal, M. K., Ahmad, P., & Altaf, M. A. (2024). From stress to success: Silicon and nano-silicon strategies for enhancing fruit yield and quality in cadmium-exposed trees. *South African Journal of Botany*, *167*, 441–447. DOI: 10.1016/j.sajb.2024.02.042

Ahmed, M., Ara, N., Aman, F., Farooq, A., Asghar, M., Muhammad, N., Ahmad, Z., & Bibi, R. (2024). Effect of iron and zinc on growth and yield of chili (*Capsicum annuum* L.). *Journal of Agriculture and Education Research*, 2(2), 1–9.

Ali, M. S., & Baek, K.-H. (2020). Jasmonic acid signaling pathway in response to abiotic stresses in plants. *International Journal of Molecular Sciences*, *21*(2), 621. DOI: 10.3390/ijms21020621

Alikhani, O., & Abbaspour, H. (2019). Effects of methyl jasmonate and cadmium on growth traits, cadmium transport and accumulation, and allene-oxide cyclase gene expression in wheat seedlings. *Revista de Agricultura Neotropical*, *6*(3), 20–29. DOI: 10.32404/rean.v6i3.3322

Altaf, M. A., Naz, S., Kumar, R., Sardar, H., Nawaz, M. A., Kumar, A., Lal, P., Ahmad, R., Hayat, F., Wani, M. A., Tiwari, R. K., & Lal, M. K. (2023). Unraveling the mechanisms of cadmium toxicity in horticultural plants: Implications for plant health. *South African Journal of Botany*, *163*, 433–442. DOI: 10.1016/j.sajb.2023.10.064

Altaf, M. A., Shahid, R., Naz, S., Ahmad, R., Manzoor, M. A., Alsahli, A. A., Altaf, M. M., & Ahmad, P. (2024). Vanadium toxicity was alleviated by supplementation of silicon in tomato seedlings: Upregulating antioxidative enzymes and glyoxalase system. *Plant Physiology and Biochemistry : PPB*, *213*, 108809. DOI: 10.1016/j.plaphy.2024.108809

Angon, P. B., Islam, M. S., Das, A., Anjum, N., Poudel, A., & Suchi, S. A. (2024). Sources, effects and present perspectives of heavy metals contamination: Soil, plants and human food chain. [Note: Article number added where typical; adjust if exact ID known]. *Heliyon*, *10*(7), e28725. DOI: 10.1016/j.heliyon.2024.e28357

Attia, H., & Alamer, K. H. (2024). Supplementation of jasmonic acid mitigates the damaging effects of arsenic stress on growth, photosynthesis and nitrogen metabolism in rice. *Rice (New York, N.Y.)*, *17*(1), 31. DOI: 10.1186/s12284-024-00709-9

Avanci, N. C., Luche, D. D., Goldman, G. H., & Goldman, M. H. D. S. (2010). Jasmonates are phytohormones with multiple functions, including plant defense and reproduction. *Genetics and Molecular Research*, *9*(1), 484–505. DOI: 10.4238/vol9-1gmr754

Bücker-Neto, L., Paiva, A. L. S., Machado, R. D., Arenhart, R. A., & Margis-Pinheiro, M. (2017). Interactions between plant hormones and heavy metals responses. *Genetics and Molecular Biology*, *40*(1, Suppl 1), 373–386. DOI: 10.1590/1678-4685-gmb-2016-0087

Chen, X., Jiang, W., Tong, T., Chen, G., Zeng, F., Jang, S., Gao, W., Li, Z., Mak, M., Deng, F., & Chen, Z. H. (2021). Molecular interaction and evolution of jasmonate signaling with transport and detoxification of heavy metals and metalloids in plants. *Frontiers in Plant Science*, *12*, 665842. DOI: 10.3389/fpls.2021.665842

Cheong, J. J., & Choi, D. Y. (2003). Methyl jasmonate as a vital substance in plants. *Trends in Genetics : TIG, 19*(7), 409–413. DOI: 10.1016/S0168-9525(03)00138-0

Ekinci, M., Shams, M., Turan, M., Ucar, S., Yaprak, E., Yuksel, E. A., Aydin, M., Ilhan, E., Agar, G., Ercisli, S., & Yildirim, E. (2024). Chitosan mitigated the adverse effect of Cd by regulating antioxidant activities, hormones, and organic acids contents in pepper (*Capsicum annuum* L.). *Heliyon, 10*(17), Article e366xx. (Use exact article ID if available)

El-Mogy, M. M., Jan, M. F., Naz, I., Ahmad, I., Ahmad, R., & Altaf, M. T. (2025). Nanoparticle innovations for mitigating metal toxicity in plants. [Volume/issue/pages as provided]. *Phyton, 94*(3), xxx–xxx. DOI: 10.32604/phyton.2025.063763

El-Saadony, M. T., Desoky, E. S. M., El-Tarabily, K. A., AbuQamar, S. F., & Saad, A. M. (2024). Exploiting the role of plant growth promoting rhizobacteria in reducing heavy metal toxicity of pepper (*Capsicum annuum* L.). *Environmental Science and Pollution Research International, 31*(18), 27465–27484. DOI: 10.1007/s11356-024-32874-1

Farooq, M. A., Gill, R. A., Islam, F., Ali, B., Liu, H., Xu, J., He, S., & Zhou, W. (2016). Methyl jasmonate regulates antioxidant defense and suppresses arsenic uptake in *Brassica napus* L. *Frontiers in Plant Science, 7*, 468. DOI: 10.3389/fpls.2016.00468

Hasan, N., Choudhary, S., Naaz, N., & Sharma, N. (2021). The mechanism of heavy metal elements in various biological process and its deteriorate effects on the productivity of different crop plants. *International Journal of Waste Resources, 11*, 403.

He, Y., Fukushige, H., Hildebrand, D. F., & Gan, S. (2002). Evidence supporting a role of jasmonic acid in Arabidopsis leaf senescence. *Plant Physiology, 128*(3), 876–884. DOI: 10.1104/pp.010843

Hewedy, O. A., Elsheery, N. I., Karkour, A. M., Elhamouly, N., Arafa, R. A., Mahmoud, G. A. E., Dawood, M. F. A., Hussein, W. E., Mansour, A., Amin, D. H., Allakhverdiev, S. I., Zivcak, M., & Brestic, M. (2023). Jasmonic acid regulates plant development and orchestrates stress response during tough times. *Environmental and Experimental Botany*, *208*, 105260. DOI: 10.1016/j.envexpbot.2023.105260

Hu, Z., Fu, Q., Zheng, J., Zhang, A., & Wang, H. (2020). Transcriptomic and metabolomic analyses reveal that melatonin promotes melon root development under copper stress by inhibiting jasmonic acid biosynthesis. *Horticulture Research*, *7*(1), 59. DOI: 10.1038/s41438-020-0293-5

Huang, H., Liu, B., Liu, L., & Song, S. (2017). Jasmonate action in plant growth and development. *Journal of Experimental Botany*, *68*(6), 1349–1359. DOI: 10.1093/jxb/erw495

Huihui, Z., Xin, L., Zisong, X., Yue, W., Zhiyuan, T., Meijun, A., Yuehui, Z., Wenxu, Z., Nan, X., & Guangyu, S. (2020). Toxic effects of heavy metals Pb and Cd on mulberry (*Morus alba* L.) seedling leaves: Photosynthetic function and reactive oxygen species (ROS) metabolism responses. *Ecotoxicology and Environmental Safety*, *195*, 110469. DOI: 10.1016/j.ecoenv.2020.110469

Jaiswal, A., Verma, A., & Jaiswal, P. (2018). Detrimental effects of heavy metals in soil, plants, and aquatic ecosystems and in humans. [Pages standardized where possible]. *Journal of Environmental Pathology, Toxicology and Oncology : Official Organ of the International Society for Environmental Toxicology and Cancer*, *37*(3), 183–197. DOI: 10.1615/JEnvironPatholToxicolOncol.2018025348

Jang, G., Yoon, Y., & Choi, Y. D. (2020). Crosstalk with jasmonic acid integrates multiple responses in plant development. *International Journal of Molecular Sciences*, *21*(1), 305. DOI: 10.3390/ijms21010305

Kaushik, S., Ranjan, A., Singh, A. K., & Sirhindi, G. (2024). Methyl jasmonate reduces cadmium toxicity by enhancing phenol and flavonoid metabolism and activating the antioxidant defense system in pigeon pea (*Cajanus cajan*). *Chemosphere*, *346*, 140681. DOI: 10.1016/j.chemosphere.2023.140681

Keramat, B., Kalantari, K. M., & Arvin, M. J. (2009). Effects of methyl jasmonate in regulating cadmium induced oxidative stress in soybean plant (*Glycine max* L.). *African Journal of Microbiological Research, 3*(5), 240–244.

Khan, A., Khan, S., Khan, M. A., Qamar, Z., & Waqas, M. (2015). The uptake and bioaccumulation of heavy metals by food plants, their effects on plants nutrients, and associated health risk: A review. *Environmental Science and Pollution Research International, 22*(18), 13772–13799. DOI: 10.1007/s11356-015-4881-0

Khan, M., Nawaz, N., Ali, I., Azam, M., Rizwan, M., Ahmad, P., & Ali, S. (2019). Regulation of photosynthesis under metal stress. In *Photosynthesis, productivity and environmental stress* (pp. 95–105). DOI: 10.1002/9781119501800.ch6

Lei, G. J., Sun, L., Sun, Y., Zhu, X. F., Li, G. X., & Zheng, S. J. (2020). Jasmonic acid alleviates cadmium toxicity in Arabidopsis via suppression of cadmium uptake and translocation. *Journal of Integrative Plant Biology, 62*(2), 218–227. DOI: 10.1111/jipb.12801

Li, D., Zhou, C., Ma, J., Wu, Y., Kang, L., An, Q., Zhang, J., Deng, K., Li, J. Q., & Pan, C. (2021). Nanoselenium transformation and inhibition of cadmium accumulation by regulating the lignin biosynthetic pathway and plant hormone signal transduction in pepper plants. *Journal of Nanobiotechnology, 19*(1), 316. DOI: 10.1186/s12951-021-01061-6

Li, Y., Sun, H., Liu, Z., Chu, Y., Huang, Y., & Bao, Q. (2024). Foliar jasmonic acid application reduces Cd and As accumulations in rice grains by regulating physiological, biochemical, and ROS scavenging attributes. *Environmental Technology & Innovation, 34*, 103596. DOI: 10.1016/j.eti.2024.103596

Liu, H., & Timko, M. P. (2021). Jasmonic acid signaling and molecular crosstalk with other phytohormones. *International Journal of Molecular Sciences, 22*(6), 2914. DOI: 10.3390/ijms22062914

Manzoor, M., Korolev, K., & Ahmad, R. (2025). Zinc oxide nanoparticles: Abiotic stress tolerance in fruit crops focusing on sustainable production. *Phyton, 94*(5), 1401–1418. DOI: 10.32604/phyton.2025.063930

Mehdizadeh, L., Moghaddam, M., Ganjeali, A., & Mahmoodi Sourestani, M. (2024). Changes in growth, essential oil composition and biochemical traits of peppermint in response to coapplication of zinc and methyl jasmonate in soilless culture. *Journal of Plant Nutrition*, *47*(16), 2732–2752. DOI: 10.1080/01904167.2024.2369063

Mir, M. A., John, R., Alyemeni, M. N., Alam, P., & Ahmad, P. (2018). Jasmonic acid ameliorates alkaline stress by improving growth performance, ascorbate glutathione cycle and glyoxylase system in maize seedlings. *Scientific Reports*, *8*(1), 2831. DOI: 10.1038/s41598-018-21097-3

Mohamed, H. I., & Latif, H. H. (2017). Improvement of drought tolerance of soybean plants by using methyl jasmonate. *Physiology and Molecular Biology of Plants : an International Journal of Functional Plant Biology*, *23*(3), 545–556. DOI: 10.1007/s12298-017-0451-x

Mousavi, S. R., Niknejad, Y., Fallah, H., & Tari, D. B. (2020). Methyl jasmonate alleviates arsenic toxicity in rice. *Plant Cell Reports*, *39*(8), 1041–1060. DOI: 10.1007/s00299-020-02547-7

Muhammad, H. M. D., Naz, S., Bibi, Z., Ahmed, M. A., Stanciu, A. S., & Ahmad, R. (2025). Nanoparticles and phytohormonal synergy in plants: Sustainable agriculture approach. *Phyton*, *94*(9), 2631–2648. DOI: 10.32604/phyton.2025.069474

Muhammad, H. M. D., Naz, S., Lal, M. K., Tiwari, R. K., Ahmad, R., Nawaz, M. A., Das, R., & Altaf, M. A. (2024). Melatonin in business with abiotic stresses in vegetable crops. *Scientia Horticulturae*, *324*, 112594. DOI: 10.1016/j.scienta.2023.112594

Naz, S., Anjum, M. A., Sadiq, B., Ahmad, R., Altaf, M. A., El-Sheikh, M. A., & Shakoor, A. (2022). Purification of sewage wastewater though sand column filter for lessening of heavy metals accumulation in lettuce, carrot, and cauliflower. *Water (Basel)*, *14*(22), 3770. DOI: 10.3390/w14223770

Nazir, F., Jahan, B., Iqbal, N., Rajurkar, A. B., Siddiqui, M. H., & Khan, M. I. R. (2023). Methyl jasmonate influences ethylene formation, defense systems, nutrient homeostasis and carbohydrate metabolism to alleviate arsenic-induced stress in rice (*Oryza sativa*). *Plant Physiology and Biochemistry : PPB*, *202*, 107990. DOI: 10.1016/j.plaphy.2023.107990

Nguyen, T. H., Goossens, A., & Lacchini, E. (2022). Jasmonate: A hormone of primary importance for plant metabolism. *Current Opinion in Plant Biology*, *67*, 102197. DOI: 10.1016/j.pbi.2022.102197

Niu, L., Zhao, H., Tang, Y., Zhu, B., Zhao, Y., Wang, Q., & Yu, J. (2025). Methyl jasmonate was involved in hydrogen sulfide-alleviated cadmium stress in cucumber plants through ROS homeostasis and chlorophyll metabolism. *International Journal of Molecular Sciences*, *26*(2), 475. DOI: 10.3390/ijms26020475

Pascual, L. S., Mittler, R., Sinha, R., Peláez-Vico, M. Á., López-Climent, M. F., Vives-Peris, V., Gómez-Cadenas, A., & Zandalinas, S. I. (2023). Jasmonic acid is required for tomato acclimation to multifactorial stress combination. *Environmental and Experimental Botany*, *213*, 105425. DOI: 10.1016/j.envexpbot.2023.105425

Per, T. S., Khan, N. A., Masood, A., & Fatma, M. (2016). Methyl jasmonate alleviates cadmium-induced photosynthetic damages through increased S-assimilation and glutathione production in mustard. *Frontiers in Plant Science*, *7*, 1933. DOI: 10.3389/fpls.2016.01933

Rai, P. K., Lee, S. S., Zhang, M., Tsang, Y. F., & Kim, K. H. (2019). Heavy metals in food crops: Health risks, fate, mechanisms, and management. *Environment International*, *125*, 365–385. DOI: 10.1016/j.envint.2019.01.067

Rashid, A., Schutte, B. J., Ulery, A., Deyholos, M. K., Sanogo, S., Lehnhoff, E. A., & Beck, L. (2023). Heavy metal contamination in agricultural soil: Environmental pollutants affecting crop health. *Agronomy (Basel)*, *13*(6), 1521. DOI: 10.3390/agronomy13061521

Raza, A., Charagh, S., Zahid, Z., Mubarik, M. S., Javed, R., Siddiqui, M. H., & Hasanuzzaman, M. (2021). Jasmonic acid: A key frontier in conferring abiotic stress tolerance in plants. *Plant Cell Reports*, *40*(8), 1513–1541. DOI: 10.1007/s00299-020-02614-z

Rehman, M., Saeed, M. S., Fan, X., Salam, A., Munir, R., Yasin, M. U., Khan, A. R., Muhammad, S., Ali, B., Ali, I., Khan, J., & Gan, Y. (2023). The multifaceted role of jasmonic acid in plant stress mitigation: An overview. *Plants*, *12*(23), 3982. DOI: 10.3390/plants12233982

Repkina, N., Murzina, S. A., Voronin, V. P., & Kaznina, N. (2023). Does methyl jasmonate effectively protect plants under heavy metal contamination? Fatty acid content in wheat leaves exposed to cadmium with or without exogenous methyl jasmonate application. *Biomolecules*, *13*(4), 582. DOI: 10.3390/biom13040582

Riyazuddin, R., Nisha, N., Ejaz, B., Khan, M. I. R., Kumar, M., Ramteke, P. W., & Gupta, R. (2021). A comprehensive review on the heavy metal toxicity and sequestration in plants. *Biomolecules*, *12*(1), 43. DOI: 10.3390/biom12010043

Saini, S., Kaur, N., & Pati, P. K. (2021). Phytohormones: Key players in the modulation of heavy metal stress tolerance in plants. *Ecotoxicology and Environmental Safety*, *223*, 112578. DOI: 10.1016/j.ecoenv.2021.112578

Sharma, A., Shahzad, B., Kumar, V., Kohli, S. K., Sidhu, G. P. S., Bali, A. S., Handa, N., Kapoor, D., Bhardwaj, R., & Zheng, B. (2019). Phytohormones regulate accumulation of osmolytes under abiotic stress. *Biomolecules*, *9*(7), 285. DOI: 10.3390/biom9070285

Sharma, R. K., & Agrawal, M. (2005). Biological effects of heavy metals: An overview. *Journal of Environmental Biology*, *26*(2), 301–313.

Shi, J., Li, J., Pan, Y., Zhao, M., Zhang, R., Xue, Y., & Liu, Y. (2024). The physiological response mechanism of peanut leaves under Al stress. *Plants*, *13*(12), 1606. DOI: 10.3390/plants13121606

Shikha, D., Jakhar, P., & Satbhai, S. B. (2023). Role of jasmonate signaling in the regulation of plant responses to nutrient deficiency. *Journal of Experimental Botany*, *74*(4), 1221–1243. DOI: 10.1093/jxb/erac387

Siddiqi, K. S., & Husen, A. (2019). Plant response to jasmonates: Current developments and their role in changing environment. *Bulletin of the National Research Center*, *43*(1), 1. DOI: 10.1186/s42269-019-0195-6

Sohn, S. I., Pandian, S., Rakkammal, K., Largia, M. J. V., Thamilarasan, S. K., Balaji, S., Zoclanclounon, Y. A. B., Shilpha, J., & Ramesh, M. (2022). Jasmonates in plant growth and development and elicitation of secondary metabolites: An updated overview. *Frontiers in Plant Science*, *13*, 942789. DOI: 10.3389/fpls.2022.942789

Song, C., Bibi, Z., Ali, S., & Ahmad, R. (2026). Silicon mitigates aluminum toxicity in lettuce by improving photosynthetic pigments and antioxidant defense system. *Biometals : An International Journal on the Role of Metal Ions in Biology, Biochemistry, and Medicine*. Advance online publication. DOI: 10.1007/s10534-026-00819-5

Sood, M. (2023). Jasmonates: "The master switch" for regulation of developmental and stress responses in plants. *Journal of Plant Growth Regulation*, *42*(8), 5247–5265. DOI: 10.1007/s00344-023-11047-3

Tu, C. Y., Wang, H. Y., Jiang, Q., Gao, Y. Q., Shen, R. F., & Zhu, X. F. (2025). Jasmonic acid enhances aluminum tolerance of rice by regulating detoxification genes. *Plant Physiology and Biochemistry : PPB*, *228*, 110257. DOI: 10.1016/j.plaphy.2025.110257

Ulloa-Inostroza, E. M., Alberdi, M., Ivanov, A. G., & Reyes-Díaz, M. (2019). Protective effect of methyl jasmonate on photosynthetic performance and its association with antioxidants in contrasting aluminum-resistant blueberry cultivars exposed to aluminum. *Journal of Soil Science and Plant Nutrition*, *19*(1), 203–216. DOI: 10.1007/s42729-019-0006-z

Verma, G., Srivastava, D., Narayan, S., Shirke, P. A., & Chakrabarty, D. (2020). Exogenous application of methyl jasmonate alleviates arsenic toxicity by modulating its uptake and translocation in rice (*Oryza sativa* L.). *Ecotoxicology and Environmental Safety*, *201*, 110735. DOI: 10.1016/j.ecoenv.2020.110735

Wang, J., Song, L., Gong, X., Xu, J., & Li, M. (2020). Functions of jasmonic acid in plant regulation and response to abiotic stress. *International Journal of Molecular Sciences*, *21*(4), 1446. DOI: 10.3390/ijms21041446

Wang, X., Cui, J., Li, X., Wang, G., Zheng, G., Li, Y., Li, X., Ma, L., & Wang, L. (2022). Comparative study on growth, physiological and biotic stress resistance of cucumber (*Cucumis sativus* L.) by exogenous application of jasmonic acid and methyl jasmonate under greenhouse conditions. ... (Complete journal details as needed)

Wasternack, C., & Hause, B. (2013). Jasmonates: Biosynthesis, perception, signal transduction and action in plant stress response, growth and development. An update to the 2007 review in *Annals of Botany. Annals of Botany*, *111*(6), 1021–1058. DOI: 10.1093/aob/mct067

Wei, T., Li, X., Yashir, N., Li, H., Sun, Y., Hua, L., Ren, X., & Guo, J. (2021). Effect of exogenous silicon and methyl jasmonate on the alleviation of cadmium-induced phytotoxicity in tomato plants. *Environmental Science and Pollution Research International*, *28*(37), 51854–51864. DOI: 10.1007/s11356-021-14252-3

Wu, B., Zeng, Z., Wu, X., Li, Y., Wang, F., Yang, J., & Li, X. (2022). Jasmonic acid negatively regulation of root growth in Japonica rice (*Oryza sativa* L.) under cadmium treatment. *Plant Growth Regulation*, *98*(3), 651–667. DOI: 10.1007/s10725-022-00897-8

Yan, Z., Zhang, W., Chen, J., & Li, X. (2015). Methyl jasmonate alleviates cadmium toxicity in *Solanum nigrum* by regulating metal uptake and antioxidative capacity. *Biologia Plantarum*, *59*(2), 373–381. DOI: 10.1007/s10535-015-0491-4

Yu, H., Li, W., Liu, X., Song, Q., Li, J., & Xu, J. (2024). Physiological and molecular bases of the nickel toxicity responses in tomato. *Stress Biology*, *4*(1), 25. DOI: 10.1007/s44154-024-00162-0

Yu, X., Zhang, W., Zhang, Y., Zhang, X., Lang, D., & Zhang, X. (2018). The roles of methyl jasmonate to stress in plants. *Functional Plant Biology : FPB*, *46*(3), 197–212. DOI: 10.1071/FP18106

Zhang, C., Huang, R., Zhan, N., & Qin, L. (2023). Methyl jasmonate and selenium synergistically mitigative cadmium toxicity in hot pepper (*Capsicum annuum* L.) plants by improving antioxidase activities and reducing Cd accumulation. *Environmental Science and Pollution Research International*, *30*(34), 82458–82469. DOI: 10.1007/s11356-023-28273-7

Zhang, D., Wang, H., Zhang, Y., Su, Z., Hu, T., Liu, J., Ding, Q., Niu, N., & Ma, L. (2024). Methyl jasmonate enhances the safe production ability of Cd-stressed wheat by regulating the antioxidant capacity, Cd absorption, and distribution in wheat. *Plant Physiology and Biochemistry : PPB*, *212*, 108788. DOI: 10.1016/j.plaphy.2024.108788

Zhou, M. L., Yang, X. B., Zhang, Q., Zhou, M., Zhao, E. Z., Tang, Y. X., Zhu, X. M., Shao, J. R., & Wu, Y. M. (2013). Induction of annexin by heavy metals and jasmonic acid in *Zea mays*. *Functional & Integrative Genomics*, *13*(2), 241–251. DOI: 10.1007/s10142-013-0316-5

Chapter 6

# Salicylic Acid Modulates Plant Tolerance Against Metal Toxicity:

## From Morphological Responses to Biochemical Mechanisms

**Zarina Bibi**
http://orcid.org/0000-0002-9016-1697
*The University of Agriculture, Dera Ismail Khan, Pakistan*

**Mohamed A. A. Ahmed**
*Alexandria University, Egypt*

## ABSTRACT

*Metals at optimum level are necessary for plants and human heaths. However, their excess is dangerous for living organisms. Heavy metals can alter the permeability of mitochondrial membranes and limit the activity of reactive oxygen species (ROS) clearance enzymes in the cellular antioxidant system. However, they cause plants to produce excessive amounts of ROS. The buildup of environmental contaminants and oxidative stress in exposed species is caused by the interaction of ROS with cellular membranes, heavy metal-induced interactions with various macromolecules either directly or indirectly, and signaling pathways.*

DOI: 10.4018/979-8-3693-9899-9.ch006

*Numerous pathogenic processes, including energy depletion, excessive ROS production, mitochondrial respiratory chain damage, decoupling of oxidative phosphorylation, and mitochondrial death, are obstructed by the heavy metal–ROS–cell signaling axis. Salicylic acid had excellent potential mitigates adverse effects of metal toxicity by regulation of plant growth, photosynthetic pigments, and antioxidant defense system in plants.*

## INTRODUCTION

Heavy metal contaminants are released into the environment as a result of anthropogenic activities, including urbanization and industrialization, that disrupt plant physiology and development through phytotoxicity (Zulfiqar et al., 2022). Soils are polluted by metals due to several factors, such as industrial waste, bombardments of agrochemicals, and rapid urbanization. The secondary metabolites also contribute to the regulation of reactive oxygen species (ROS). For instance, the main secondary metabolites that combat ROS and prevent oxidative stress in plants include vitamins, terpenes, and polyphenols (Rai et al., 2021). The higher accumulation of metals in plants later affects the physicochemical activities of plants (Salifu et al., 2024). Plants have developed defense mechanisms against heavy metal-induced oxidative injury with increased metal extraction from roots, prevention of metal entry into the plant, chelation by organic compounds, sequestration of metals in vacuoles, and metal binding in the cell wall to prevent entry (Al-Obaidi et al., 2024). ROS are also transformed into less harmful compounds. Different concentrations of metals are found naturally in the Earth's crust, and metabolic systems find it difficult to convert them into less harmful molecules. They have detrimental effects on the ecosystem and human health when they are found in soil, where they persist for long periods (Alsafran et al. 2022).

Heavy metals modulate several physiological systems, mineral nutritional status, photosynthesis and respiration rates, enzymatic processes, and numerous biochemical processes regardless of these external influences. However, increased synthesis of ROS is thought to be the most immediate effect to date when plants are exposed to metal toxicity (Rizvi et al., 2020). Metal toxicity changes the activity of the electron transport chain by interacting with different components, which mostly causes ROS production in mitochondria and chloroplasts. The membrane potential becomes unbalanced when ROS levels grow, which further causes ion channel leakage, lipid peroxidation, and macromolecule destruction (Rucińiska-Sobkowiak, 2010). Furthermore, depending on a number of variables, including the length of stress, the developmental stage, the dose or concentration of a specific heavy metal, and the plant organ involved, the toxicity of heavy metals that release ROS on subcellular organelles may change (Fryzova et al., 2017). In addition to these adverse circumstances, plants also generate ROS at a baseline rate that does not result in any harmful effects due to scavenging through various antioxidant mechanisms. Because plants absorb all of the nutrients that are accessible to them, including environmental pollutants, particularly heavy metals. Furthermore, they contribute to very little toxicity within the plant because they are dispersed throughout many tissues (Maleki et al., 2017). Therefore, it is crucial to comprehend the mechanisms underlying plants' restricted uptake of heavy metals (Ali et al., 2019). Recently, considerable progress has been accomplished about signaling pathways formed in plants and their physiological response to heavy metal stress. The substantial advancements in our knowledge of the signaling pathways and physiological reactions of plants to metal toxicity (Shoaib and Javaid, 2021). Metal sequestration, the synthesis of non-enzymatic antioxidants, the activation of antioxidant defense mechanisms, and the control of metal absorption and transport are among the major events covered. Together, these tactics help the plant prevent oxidative stress brought on by heavy metals and shield its biological components from harm (Schutzendubel and Polle, 2002).

Heavy metal-induced soil pollution can be addressed economically and environmentally by phytoremediation, which uses plants to extract pollutants. Gaining a better understanding of the processes governing heavy metal accumulation and tolerance in plants is crucial to maximizing the efficacy of phytoremediation. Excessive production of ROS by plants exposed to heavy metals might result in oxidative stress (Nowicka, 2022). Therefore, increasing antioxidant activity is a frequently used strategy to improve heavy metal tolerance, strengthening the plant's resistance to oxidative stress. Heavy metal buildup in plants can be increased by genetic engineering. This strategy entails the introduction and overexpression of genes linked to heavy metal absorption, translocation, and sequestration (Sytar et al., 2013). The use of metal chelators, which can be encouraged by genetic engineering by overexpressing genes encoding natural chelators, is another intriguing tactic. This method enhances the absorption and transport of heavy metals in plants (Juknys et al., 2012). However, phytohormones are also effective for increasing metal tolerance in plants. Among phytohormones, salicylic acid (SA) improves metal tolerance in plants. Hence, the current chapter can explore the significance of SA on metal-affected plants.

## SUBSTANTIAL CONTRIBUTION OF SA TO THE PRODUCTIVITY OF CROPS UNDER METAL TOXICITY

Metal pollution is now a major environmental issue. Metals released from various industrialization, rapid population growth, and urbanization endanger soil and water and impair living organisms due to their biomagnification through the food chain (Ercal et al., 2001). Toxic elements have been released into the environment more frequently as a result of natural processes like rock erosion and volcanic eruptions, but their concentration in the biosphere has increased due to increased human activities like mining, painting, and refining (Yadav, 2010). The biotic and abiotic elements of the ecosystem are seriously threatened by metal-induced soil contamination. Plant development, biomass, photosynthesis, crop production, and plant quality may all be negatively impacted by the increased uptake of metals in soil (Rajewska et al., 2016). The topsoil is the most active area of the soil and collects a significant number of hazardous metals, which is quite concerning for the environment (Kacienė et al., 2017).

Plant cell metabolism is adversely affected by the elevated level of heavy metal buildup in plant organs (Goncharuk and Zagoskina, 2023). High concentrations of hazardous chemicals naturally impact the various physiological processes in plants, including respiration, photosynthesis, protein metabolism, and morphogenesis (Seneviratne et al., 2019). For instance, after treating rice seedlings with cadmium, Mediouni et al. (2006) observed altered morphogenesis and reduced root growth. Many plant species displayed suppression of photosynthesis when exposed to cadmium

(Piotto et al., 2018). Similarly, Sun et al. (2023) observed the impact of lead on the morphology and metabolism of plants and discovered that the plant's growth was impacted by the metal's rising concentration. At greater lead concentrations, the plant also displayed severe chlorosis, necrosis, and decreased chlorophyll and protein contents (Haouari et al., 2012).

The remediation of metal-contaminated soil ecosystems is essential for their protection and restoration. Characterization of the soil sheds light on the bioavailability and speciation of metal toxicity. Considering the causes of contamination, its chemical characteristics, and the threats to the environment and human health are necessary for remediation operations. To manage contaminated sites economically while maintaining the health of the people and ecology, risk assessment is an essential technique (Pant et al., 2014). Metal-contaminated soils can be remedied using a variety of strategies, such as physical, chemical, and biological processes. Containment and excavation are examples of physical approaches used to physically remove or isolate contaminated soil (Rani et al., 2014). Chemical treatments change the pollutants' chemical characteristics by using ion exchange and precipitation. Heavy metals are extracted and accumulated from the soil using biological techniques like phytoremediation, which uses metal-accumulating plants (Wani et al., 2023). The type and concentration of the heavy metal, the state of the environment, and the required degree of remediation all influence the choice of remediation technique. Techniques including immobilization, soil washing, and phytoremediation are widely recognized as some of the best and most proven methods for cleaning up heavy metal-contaminated areas (Yordanova et al., 2009). While soil cleaning involves the elimination of pollutants using physicochemical methods, immobilization attempts to decrease metal mobility by changing their chemical structure. Utilizing specific plants' capacity to absorb environmental conditions and the required degree of remediation, phytoremediation takes advantage of these properties (Noman and Aqeel, 2017).

Different management practices, such as immobilization, soil washing, and phytoremediation, are widely recognized as some of the best and most proven methods for cleaning up heavy metal-contaminated areas. While soil cleaning entails eliminating pollutants using chemical or physical means, immobilization attempts to decrease heavy metal mobility by changing their chemical form. Phytoremediation is also an effective strategy for the removal of pollutants from the soil (Javaid et al., 2020). However, SA had an excellent contribution to the productivity of crops under metal toxicity (Table 1). SA reduces the adverse effects of metal toxicity in plants by altering biochemical changes (Figure 1).

*Table 1. Substantial contribution of salicylic acid in productivity of crops under metal toxicity*

| Crops name | Impact of salicylic acid | References |
|---|---|---|
| Tomato | Shields cellular structures and regulates photosynthesis | Sun et al. (2023) |
| Maize | Sustains chlorophyll content, regulates stomatal conductance, and improves biomass | Gondor et al. (2016) |
| Barley | Regulate transporters and chelation of toxic ions. Reduces metal toxicity.<br>Enhancement of growth traits | Song et al. (2014) |
| Wheat | Encourages proline and glycine betaine synthesis.<br>Preserves osmotic balance and turgor pressure under metal toxicity | Rizvi et al. (2020) |
| Rice | Reduces lipid peroxidation in plants<br>Sustains cell viability and function | Wang et al. (2021) |
| Sweet potato | Interacts with ethylene, and auxin for synchronized stress reaction<br>Boosts adaptability and reproductive achievement | Kumar et al. (2022) |
| Cucumber | Regulates metabolic functions<br>Enhanced yield traits | Simek et al. (2016) |
| Peanut | Detoxifies metals and preserves metabolic system | Zong et al. (2020) |
| Peppermint | Increases nutrients assimilation and water uptake | Ahmad et al. (2018) |

*Figure 1. Salicylic acid improves crop production under metal toxicity by regulation of biochemical activities*

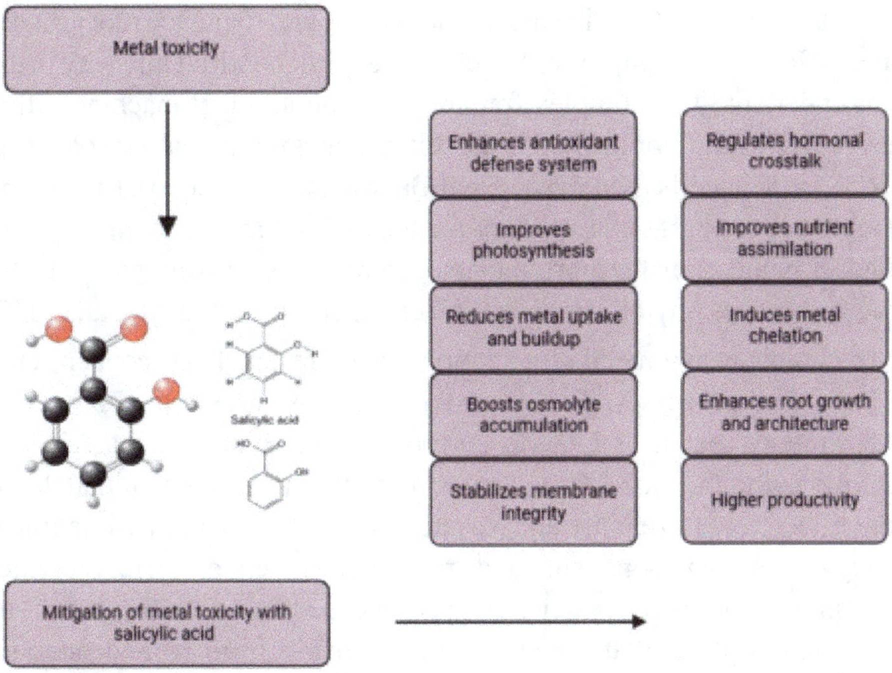


## SA ADJUSTS MORPHOLOGICAL AND BIOCHEMICAL CHANGES IN PLANTS UNDER METAL TOXICITY

The primary result of metal toxicity is the maximal generation of ROS as a result of metal toxicity interference with the photosynthetic process (Huihui et al., 2020). During electron transport in photosynthesis and respiration pathways, ROS such as hydroxyl, superoxide, and hydrogen peroxide are produced as byproducts (Küpper and Andresen, 2016). Under physiological scenarios, ROS participate in a well-coordinated and regulated network and play a signaling role in plants (Nowicka, 2022). ROS are a part of a complex regulation system in plants that also includes secondary messengers, signaling molecules, plant hormones, and ROS (Małkowski et al., 2019). ROS production, however, can build up to oxidative stress in cells when it exceeds normal levels. This can result in lipid peroxidation, macromolecular breakdown, membrane rupture, DNA breakage, and ion leakage in plants (Pardo-Hernández et al., 2021). For instance, Pb-induced ultrastructural alterations in wheat roots were found to reduce root growth, generate ROS, and compromise nuclear and mitochondrial integrity in the plant under study. A complex framework of antioxidant machinery that preserves ROS equilibrium within the cell regulates the increased production of ROS in plant cells (Kalaivanan and Ganeshamurthy, 2016). Enzymatic antioxidants are part of a well-coordinated and optimized defense system found in plants. Furthermore, plants' defense mechanism against metal toxicity involves phytohormones (Shu et al., 2012; El-Mogy et al., 2025)

To combat heavy metal remediation, recent research has concentrated on integrated processes. The electrokinetic-geosynthetic method is one such strategy that uses electric current and geosynthetic materials to improve pollutant mobility and remove metals from contaminated soil more effectively (Greičiūtė and Vasarevičius, 2007). Another integrated strategy uses microorganisms and transparent reactive barriers to filter contaminants out when water passes across the treated barrier area. Microbes and/or plants that can absorb heavy metals from groundwater are included in these barriers (Abdu et al., 2017; Manzoor et al., 2025). The use of biotechnological techniques to treat contaminated sediments is also becoming more popular. Given the serious health effects of heavy metal exposure, such as infertility problems and genetic, epigenetic, and biochemical changes, it is essential to comprehend the pathways that are activated by metal exposure to create effective remediation solutions. Because heavy metal-contaminated sites are complicated and distinctive, customized remediation solutions are required. However, substantial decreases in the levels of heavy metals in the environment depend on government organizations enforcing regulations strictly and taking firm action against companies that release harmful pollutants into the environment (Shukla et al., 2019). Salicylic acid modulates morphological and biochemical activities in plants under metal toxicity (Table 2).

*Table 2. Salicylic acid modulates morphological and biochemical activities in plants under metal toxicity*

| Effect of metal toxicity | Plant response to salicylic acid | Reference |
|---|---|---|
| Root growth inhibition and reduced shoot length | Improves root architecture with longer roots, higher biomass and the maximum number of lateral roots | Alamer and Fayez (2020) |
| Stunted growth and shortened internodes | Promotes shoot elongation and higher stem thickness | Bahrami et al. (2016) |
| Decreased leaf area, chlorosis, and necrosis | Boosts photosynthetic pigments | Gupta and Seth (2021) |
| Higher lipid peroxidation and electrolyte leakage | Enhances membrane integrity | Wang et al. (2021) |
| Decrease in chlorophyll content and carotenoids | Restores pigment levels, sustains photosynthesis under metal toxicity | Pardo-Hernández et al. (2021) |
| Disrupted proline and soluble sugar synthesis | Higher accumulation of proline, total sugars, and glycine betaine | Yordanova et al. (2009) |
| Limits metal uptake and encourage chelation | Lowers internal metal toxicity | Haouari et al. (2012) |
| Inhibited minerals uptake | Increases nutrient absorption and translocation | Kacienė et al. (2017) |
| Reduced vegetative and reproductive growth | Enhances overall plant production | Juknys et al. (2012) |

## SA REDUCES THE TOXIC SUBSTANCES PRODUCED IN PLANTS AGAINST METAL TOXICITY

One of the primary soil pollutants limiting crop productivity worldwide is metal toxicity. A decrease in production results from plants growing in metal-contaminated soils exhibiting limited growth and development (Jyothsna and Murthy, 2016). Higher soil concentrations of metal toxicity cause excessive absorption and affect the metabolic processes of plants. Metal toxicity changes the shape and structure of roots and shoots as they develop (Popova et al., 2008). SA mitigates the adverse effects of metal toxicity in several crops, such as wheat (Gil’vanova et al., 2012), barley

(Song et al., 2014), cucumber (Simek et al., 2016), rice (Wang et al., 2021), and Maize (Sofy et al., 2020). Plants have complex physiological and cellular defensive mechanisms, such as the ability to store and transport metals, to protect themselves from metal toxicity. Furthermore, plants have developed efficient techniques to use signaling pathways to induce the formation of metal transporters (Kohli et al., 2017). Plants can either prevent metal toxicity from being absorbed from the soil or initiate the detoxification process, which includes generating antioxidants to scavenge reactive oxygen species, to deal with metal toxicity (Gondor et al., 2016). Cadmium treatment significantly reduced the growth characteristics and resulted in oxidative damage by raising the concentrations of proline in pea seedlings. On the other hand, when applied to pea leaves exposed to cadmium (Cd) toxicity, SA significantly increased the growth attributed to the antioxidant enzyme system, chlorophyll content, and decreased proline production and electrolyte leakage. Cd buildup in a shoot was significantly decreased by SA treatment (Popova et al., 2008). Additionally, Gupta and Seth (2021) found that tomato seedling growth was changed by chromium (Cr) toxicity. However, SA efficiently controls tomato seedling growth. In tomato seedlings exposed to Cr toxicity, SA decreased oxidative damage and Cr accumulation while improving gas exchange properties, pigment concentration, and leaf photosynthesis. Watermelon seedlings' photosynthetic assimilation rate, pigment level, and seedling development were all considerably lowered under extreme boron toxicity. Furthermore, watermelon leaves' oxidative stress level rose in response to an excessive boron concentration. Meanwhile, SA treatment preserved ROS homeostasis, reduced Cr absorption from root to shoot, and successfully restored the leaf photosynthetic mechanism and carotenoid concentration (Bahrami et al., 2016). In potato leaves exposed to Cd toxicity, SA supplementation drastically reduced oxidative stress as well as Cd translocation from root to shoot. Furthermore, foliar SA application in sweet potato enhanced proline accumulation, gas exchange characteristics, and antioxidant enzymes (Kumar et al., 2022). In two peanut cultivars exposed to Cd toxicity, SA improved the antioxidant enzyme activity and Cd stress tolerance (Zong et al., 2020).

SA enhanced plant growth, photosynthetic system, antioxidant mechanism, ROS homeostasis, mineral homeostasis, and decreased the accumulation of metals from root to shoot (Kalaivanan and Ganeshamurthy, 2016). SA is an important growth regulator that effectively controls horticultural plant growth. Furthermore, exogenous SA regulated horticulture plant growth, decreased oxidative damage, and enhanced the antioxidant enzyme system and leaf photosynthetic mechanism (Krantev et al., 2008). SA pretreatment might aid in plant survival in environments contaminated with metals. SA successfully lessened the harmful effects of metals by decreasing the movement and transfer of metals from root to shoot (Popova et al., 2011; Muhammad et al., 2025). Furthermore, horticultural plants' mineral nutrient accumulation and root architecture were altered by SA application. The impact of exogenous SA treatment on strengthening and developing plant defense mechanisms is explored in this chapter. However, small concentrations of SA have a remarkable potential to withstand metal toxicity, making it a remarkable tactic to ensure the sustainability of agricultural crop production (Alamer and Fayez, 2020; Naz et al., 2022; Altaf et al., 2023).

## SA IMPROVES PLANT METABOLIC ACTIVITIES AND SIGNALING MOLECULES AGAINST METAL TOXICITY

SA had the potential to induce metal stress tolerance in plants. The tolerance mechanism is dependent on the plant's health as well as the SA concentration and application method (Ahmad et al., 2018). In addition to influencing plant growth and development, abiotic stress controls the machinery involved in DNA replication (Liu et al., 2011). The topoisomerase gene and chloroplast elongation factor are upregulated by SA use, which aids in plant adaptability to harsh environments (Singh et al., 2004). Furthermore, TLC1, a long terminal repeat retrotransposon family, is reported to be expressed *in vivo* by SA (Tapia et al., 2005). This family's expression by SA indicates that they play a part in SA-mediated signaling pathways, and they are transcriptionally activated under stressful circumstances (Salazar et al., 2007). The enhanced activity of enzymes involved in the ascorbate-glutathione (AsA-GSH) pathway is another method that SA has developed to regulate the response to metal toxicity in plants. AsA and GSH are both active redox substances that preserve cellular redox balance (Bitarishvili et al., 2023).

SA supplementation improved Cd-tolerance and raised antioxidant levels in barley (Tamás et al., 2015). This may be due to higher oxidative harms, which in turn stimulate the activity of antioxidant enzymes that lower the level of ROS in plants' cells (Huang et al., 2006). Through interactions with other plant growth regulators, SA regulates various plant responses in both ideal and stressful environments. Under both ideal and stressful conditions, the relationship between SA and various other hormones was also effective for tolerance of metal toxicity (Guo et al., 2019). Under stressful circumstances, the interaction between SA and hormones may have a synergistic or antagonistic effect (Ahmad et al., 2022). Similarly, Tamás et al. (2015) also explored that SA controlled the reduction of Cd-induced auxin-mediated ROS generation in barley roots, hence mitigating Cd-stress. The authors assume that SA plays a significant role in the enhancement of metal tolerance in plants by improving root growth, root size, plant fresh biomass, shoot length, and several other traits relevant to growth and yield (Muhammad et al., 2024; Neyshabouri et al., 2024).

Plants that are exposed to different environmental stressors, including metal toxicity, may produce more ethylene and experience oxidative stress (Kalai et al., 2016). Either elevated expression of ethylene-responsive genes or ethylene-related biosynthetic genes is responsible for the enhanced ethylene production (Habibi, 2012). By raising GSH levels, which led to metal detoxification and scavenging ROS brought on by metal-triggered ethylene formation, exogenous SA was shown to reduce Cd stress in wheat (Kovács et al., 2014). Due to a de novo abscisic acid (ABA) production, SA supplementation raised the amount of ABA in wheat seedlings under Cd stress (Zulfiqar et al., 2024). Moreover, the protective action of SA in wheat plants is demonstrated by the endogenous ABA-controlled SA-mediated modification of the concentration of dehydrin proteins under metal toxicity (Touzout et al., 2021; Altaf et al., 2024). The detection of Cd in plants can initiate a variety of signaling pathways. Plant hormones, both endogenous and exogenous, contribute to reducing the stress caused by cadmium (Altaf et al., 2023). Understanding how phytohormones act in plants under cadmium stress has advanced significantly in recent years (Cheng et al., 2025; Song et al., 2026). Salicylic acid mitigates the adverse effects of metal toxicity in plants by activating the plant defense system (Figure 2).

*Figure 2. Salicylic acid mitigates adverse effects of metal toxicity in plants by the activation of plant defense system*

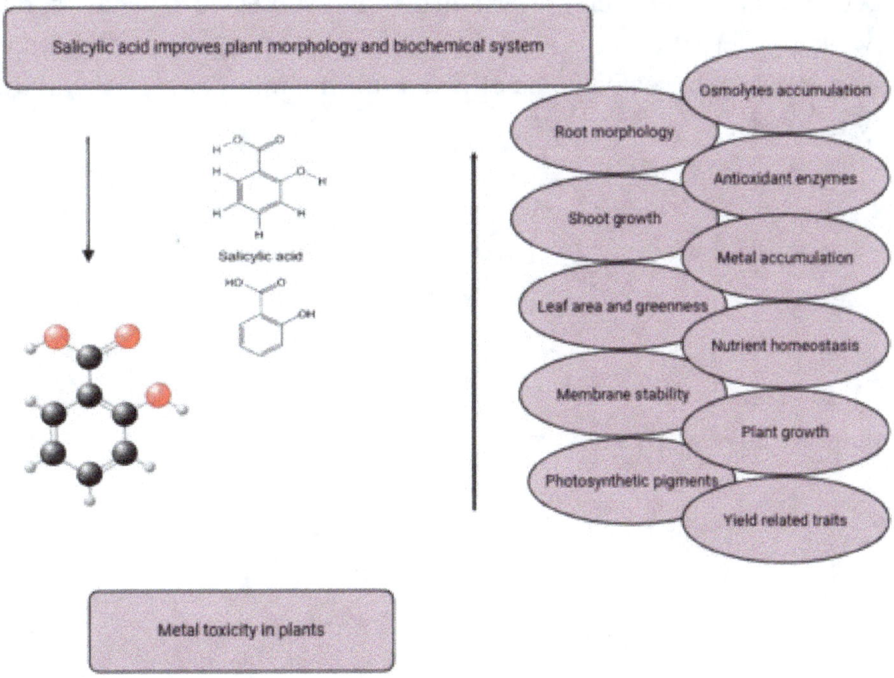


## CONCLUSION AND FUTURE OUTLOOK

One of the biggest risks to plants growing in contaminated circumstances is metal toxicity. Plants have evolved various molecular, metabolic, and physiological mechanisms to combat the negative effects of metal toxicity. These mechanisms enable plants to either avoid or manage metal stress. Salicylic acid (SA) is essential for alleviating and lowering metal toxicity in plants. Exogenous SA application improves endogenous production of SA to encourage the mitigation of the adverse effects of metal toxicity. It has been shown that both an increase in the endogenous level and an exogenous use of SA benefit plants under metal toxicity. SA controls plant metabolic functions, including the production of secondary metabolites and osmoprotectants that help plants to scavenge oxidative injury. This increases the activity of enzymes that scavenge toxic substances. Hence, in this way, SA improves plant tolerance against metal toxicity. SA could trigger the antioxidant defense system of plants subjected to metal toxicity. However, different molecular analyses, i.e., proteomics, transcriptomics, and metabolomics, can also be effective for the mitigation of metal toxicity in plants.

## REFERENCES


Abdu, N., Abdullahi, A. A., & Abdulkadir, A. (2017). Heavy metals and soil microbes. *Environmental Chemistry Letters, 15*(1), 65–84. DOI: 10.1007/s10311-016-0587-x

Ahmad, B., Jaleel, H., Sadiq, Y. A., Khan, M. M., & Shabbir, A. (2018). Response of exogenous salicylic acid on cadmium induced photosynthetic damage, antioxidant metabolism and essential oil production in peppermint. *Plant Growth Regulation, 86*(2), 273–286. DOI: 10.1007/s10725-018-0427-z

Al-Obaidi, J. R., Jamaludin, A. A., Rahman, N. A., & Ahmad-Kamil, E. I. (2024). How plants respond to heavy metal contamination: A narrative review of proteomic studies and phytoremediation applications. *Planta, 259*(5), 103. DOI: 10.1007/s00425-024-04378-2

Alamer, K. H., & Fayez, K. A. (2020). Impact of salicylic acid on the growth and physiological activities of parsley plants under lead toxicity. *Physiology and Molecular Biology of Plants : an International Journal of Functional Plant Biology, 26*(7), 1361–1373. DOI: 10.1007/s12298-020-00830-1

Ali, M. A., Fahad, S., Haider, I., Ahmed, N., Ahmad, S., Hussain, S., & Arshad, M. 2019. Oxidative stress and antioxidant defense in plants exposed to metal/metalloid toxicity. *Reactive oxygen, nitrogen and sulfur species in plants: production, metabolism, signaling and defense mechanisms*, pp.353-370.

Alsafran, M., Usman, K., Ahmed, B., Rizwan, M., Saleem, M. H., & Al Jabri, H. (2022). Understanding the phytoremediation mechanisms of potentially toxic elements: A proteomic overview of recent advances. *Frontiers in Plant Science, 13*, 881242. DOI: 10.3389/fpls.2022.881242

Altaf, M. A., Naz, S., Kumar, R., Sardar, H., Nawaz, M. A., Kumar, A., Lal, P., Ahmad, R., Hayat, F., Wani, M. A., Tiwari, R. K., & Lal, M. K. (2023). Unraveling the mechanisms of cadmium toxicity in horticultural plants: Implications for plant health. *South African Journal of Botany, 163*, 433–442. DOI: 10.1016/j.sajb.2023.10.064

Altaf, M. A., Shahid, R., Naz, S., Ahmad, R., Manzoor, M. A., Alsahli, A. A., Altaf, M. M., & Ahmad, P. (2024). Vanadium toxicity was alleviated by supplementation of silicon in tomato seedlings: Upregulating antioxidative enzymes and glyoxalase system. *Plant Physiology and Biochemistry : PPB*, *213*, 108809. DOI: 10.1016/j.plaphy.2024.108809

Bahrami, M., Heidari, M., & Ghorbani, H. (2016). Variation in antioxidant enzyme activities, growth and some physiological parameters of bitter melon (Momordica charantia) under salinity and chromium stress. *Journal of Environmental Biology*, *37*(4), 529.

Bitarishvili, S., Dikarev, A., Kazakova, E., Bondarenko, E., Prazyan, A., Makarenko, E., Babina, D., Podobed, M., & Geras'kin, S. (2023). Growth, antioxidant system, and phytohormonal status of barley cultivars contrasting in cadmium tolerance. *Environmental Science and Pollution Research International*, *30*(21), 59749–59764. DOI: 10.1007/s11356-023-26523-2

Cheng, C., Liu, J., Liu, J., Gao, Z., Yang, Y., Zhu, B., Yao, F., & Ye, Q. (2025). Selenium alleviates cadmium toxicity in pepper (Capsicum annuum L.) by reducing accumulation, enhancing stress resistance, and promoting growth. *Plants*, *14*(9), 1291. DOI: 10.3390/plants14091291

El-Mogy, M.M., Jan, M.F., Naz, I., Ahmad, I., Ahmad, R. and Altaf, M.T., 2025. Nanoparticle Innovations for Mitigating Metal Toxicity in Plants. *Phyton (0031-9457)*, *94*(3).

Ercal, N., Gurer-Orhan, H., & Aykin-Burns, N. (2001). Toxic metals and oxidative stress part I: Mechanisms involved in metal-induced oxidative damage. *Current Topics in Medicinal Chemistry*, *1*(6), 529–539. DOI: 10.2174/1568026013394831

Fryzova, R., Pohanka, M., Martinkova, P., Cihlarova, H., Brtnicky, M., Hladky, J. and Kynicky, J., 2017. Oxidative stress and heavy metals in plants. *Reviews of environmental contamination and toxicology volume 245*, pp.129-156.

Gil'vanova, I. R., Enikeev, A. R., Stepanov, S. Y., & Rakhmankulova, Z. F. (2012). Involvement of salicylic acid and nitric oxide in protective reactions of wheat under the influence of heavy metals. *Applied Biochemistry and Microbiology*, *48*(1), 90–94. DOI: 10.1134/S000368381201005X

Goncharuk, E. A., & Zagoskina, N. V. (2023). Heavy metals, their phytotoxicity, and the role of phenolic antioxidants in plant stress responses with focus on cadmium. *Molecules (Basel, Switzerland), 28*(9), 3921. DOI: 10.3390/molecules28093921

Gondor, O. K., Pál, M., Darkó, É., Janda, T., & Szalai, G. (2016). Salicylic acid and sodium salicylate alleviate cadmium toxicity to different extents in maize (Zea mays L.). *PLoS One, 11*(8), e0160157. DOI: 10.1371/journal.pone.0160157

Greičiūtė, K. and Vasarevičius, S., 2007. Decontamination of heavy-metal polluted soil BY electrokinetic remediation.

Guo, B., Liu, C., Liang, Y., Li, N., & Fu, Q. (2019). Salicylic acid signals plant defence against cadmium toxicity. *International Journal of Molecular Sciences, 20*(12), 2960. DOI: 10.3390/ijms20122960

Gupta, S., & Seth, C. S. (2021). Salicylic acid alleviates chromium (VI) toxicity by restricting its uptake, improving photosynthesis and augmenting antioxidant defense in Solanum lycopersicum L. *Physiology and Molecular Biology of Plants : an International Journal of Functional Plant Biology, 27*(11), 2651–2664. DOI: 10.1007/s12298-021-01088-x

Habibi, G. (2012). Exogenous salicylic acid alleviates oxidative damage of barley plants under drought stress. *Acta Biologica Szegediensis, 56*(1), 57–63.

Haouari, C. C., Nasraoui, A. H., Bouthour, D., Houda, M. D., Daieb, C. B., Mnai, J., & Gouia, H. (2012). Response of tomato (Solanum lycopersicon) to cadmium toxicity: Growth, element uptake, chlorophyll content and photosynthesis rate. *African Journal of Plant Science, 6*(1), 1–7.

Huang, Y., Zhang, G., Wu, F., Chen, J., & Xiao, Y. (2006). Interaction of salinity and cadmium stresses on antioxidant enzymes, sodium, and cadmium accumulation in four barley genotypes. *Journal of Plant Nutrition, 29*(12), 2215–2225. DOI: 10.1080/01904160600974155

Huihui, Z., Xin, L., Zisong, X., Yue, W., Zhiyuan, T., Meijun, A., Yuehui, Z., Wenxu, Z., Nan, X., & Guangyu, S. (2020). Toxic effects of heavy metals Pb and Cd on mulberry (Morus alba L.) seedling leaves: Photosynthetic function and reactive oxygen species (ROS) metabolism responses. *Ecotoxicology and Environmental Safety*, *195*, 110469. DOI: 10.1016/j.ecoenv.2020.110469

Javaid, S., uz Zaman, Q., Sultan, K., Riaz, U., Aslam, A., Sharif, S., ... & Ibraheem, S. (2020). 5. Heavy metals stress, mechanism and remediation techniques in rice (Oryza sativa L.): a review. Pure and Applied Biology (PAB), 9(1), 403-426.

Juknys, R., Vitkauskaitė, G., Račaitė, M., & Venclovienė, J. (2012). The impacts of heavy metals on oxidative stress and growth of spring barley. *Central European Journal of Biology*, *7*(2), 299–306.

Jyothsna, P. I., & Murthy, S. D. S. (2016). Review on role of salicylic acid in the alleviation of cadmium induced damage in plants. *Research Journal of Pharmaceutical, Biological and Chemical Sciences*, *7*(4), 2746–2751.

Kacienė, G., Juknys, R., & Januškaitienė, I. (2017). The role of oxidative stress in spring barley cross-adaptation to different heavy metals. *Archives of Agronomy and Soil Science*, *63*(8), 1037–1048. DOI: 10.1080/03650340.2016.1256474

Kalai, T., Bouthour, D., Manai, J., Bettaieb Ben Kaab, L., & Gouia, H. (2016). Salicylic acid alleviates the toxicity of cadmium on seedling growth, amylases and phosphatases activity in germinating barley seeds. *Archives of Agronomy and Soil Science*, *62*(6), 892–904. DOI: 10.1080/03650340.2015.1100295

Kalaivanan, D., & Ganeshamurthy, A. N. (2016). Mechanisms of heavy metal toxicity in plants. In *Abiotic stress physiology of horticultural crops* (pp. 85–102). Springer India. DOI: 10.1007/978-81-322-2725-0_5

Kalaivanan, D., & Ganeshamurthy, A. N. (2016). Mechanisms of heavy metal toxicity in plants. In *Abiotic stress physiology of horticultural crops* (pp. 85–102). Springer India. DOI: 10.1007/978-81-322-2725-0_5

Kohli, S. K., Handa, N., Kaur, R., Kumar, V., Khanna, K., Bakshi, P., Singh, R., Arora, S., Kaur, R., & Bhardwaj, R. (2017). Role of salicylic acid in heavy metal stress tolerance: insight into underlying mechanism. In *Salicylic acid: a multifaceted hormone* (pp. 123–144). Springer Singapore. DOI: 10.1007/978-981-10-6068-7_7

Kovács, V., Gondor, O. K., Szalai, G., Darkó, É., Majláth, I., Janda, T., & Pál, M. (2014). Synthesis and role of salicylic acid in wheat varieties with different levels of cadmium tolerance. *Journal of Hazardous Materials*, *280*, 12–19. DOI: 10.1016/j.jhazmat.2014.07.048

Krantev, A., Yordanova, R., Janda, T., Szalai, G., & Popova, L. (2008). Treatment with salicylic acid decreases the effect of cadmium on photosynthesis in maize plants. *Journal of Plant Physiology*, *165*(9), 920–931. DOI: 10.1016/j.jplph.2006.11.014

Kumar, S., Wang, M., Fahad, S., Qayyum, A., Chen, Y., & Zhu, G. (2022). Chromium induces toxicity at different phenotypic, physiological, biochemical, and ultrastructural levels in sweet potato (Ipomoea batatas L.) plants. *International Journal of Molecular Sciences*, *23*(21), 13496. DOI: 10.3390/ijms232113496

Küpper, H., & Andresen, E. (2016). Mechanisms of metal toxicity in plants. *Metallomics : Integrated Biometal Science*, *8*(3), 269–285. DOI: 10.1039/C5MT00244C

Liu, C., Guo, J., Cui, Y., Lü, T., Zhang, X., & Shi, G. (2011). Effects of cadmium and salicylic acid on growth, spectral reflectance and photosynthesis of castor bean seedlings. *Plant and Soil*, *344*(1), 131–141. DOI: 10.1007/s11104-011-0733-y

Maleki, M., Ghorbanpour, M., & Kariman, K. (2017). Physiological and antioxidative responses of medicinal plants exposed to heavy metals stress. *Plant Gene*, *11*, 247–254. DOI: 10.1016/j.plgene.2017.04.006

Małkowski, E., Sitko, K., Zieleźnik-Rusinowska, P., Gieroń, Ż., & Szopiński, M. (2019). Heavy metal toxicity: Physiological implications of metal toxicity in plants. In *Plant metallomics and functional omics: a system-wide perspective* (pp. 253–301). Springer International Publishing. DOI: 10.1007/978-3-030-19103-0_10

Manzoor, M., Korolev, K. and Ahmad, R., 2025. Zinc Oxide Nanoparticles: Abiotic Stress Tolerance in Fruit Crops Focusing on Sustainable Production. *Phyton (0031-9457), 94*(5).

Mediouni, C., Benzarti, O., Tray, B., Ghorbel, M. H., & Jemal, F. (2006). Cadmium and copper toxicity for tomato seedlings. *Agronomy for Sustainable Development, 26*(4), 227–232. DOI: 10.1051/agro:2006008

Muhammad, H.M.D., Naz, S., Bibi, Z., Ahmed, M.A., Stanciu, A.S. and Ahmad, R., 2025. Nanoparticles and Phytohormonal Synergy in Plants: Sustainable Agriculture Approach. *Phyton (0031-9457), 94*(9).

Muhammad, H. M. D., Naz, S., Lal, M. K., Tiwari, R. K., Ahmad, R., Nawaz, M. A., Das, R., & Altaf, M. A. (2024). Melatonin in business with abiotic stresses in vegetable crops. *Scientia Horticulturae, 324*, 112594. DOI: 10.1016/j.scienta.2023.112594

Naz, S., Anjum, M. A., Sadiq, B., Ahmad, R., Altaf, M. A., El-Sheikh, M. A., & Shakoor, A. (2022). Purification of sewage wastewater though sand column filter for lessening of heavy metals accumulation in lettuce, carrot, and cauliflower. *Water (Basel), 14*(22), 3770. DOI: 10.3390/w14223770

Neyshabouri, F. A., Ghotbi-Ravandi, A. A., Shariatmadari, Z., & Tohidfar, M. (2024). Cadmium toxicity promotes hormonal imbalance and induces the expression of genes involved in systemic resistances in barley. *Biometals : An International Journal on the Role of Metal Ions in Biology, Biochemistry, and Medicine, 37*(5), 1147–1160. DOI: 10.1007/s10534-024-00597-y

Noman, A., & Aqeel, M. (2017). miRNA-based heavy metal homeostasis and plant growth. *Environmental Science and Pollution Research International, 24*(11), 10068–10082. DOI: 10.1007/s11356-017-8593-5

Nowicka, B. (2022). Heavy metal–induced stress in eukaryotic algae—Mechanisms of heavy metal toxicity and tolerance with particular emphasis on oxidative stress in exposed cells and the role of antioxidant response. *Environmental Science and Pollution Research International, 29*(12), 16860–16911. DOI: 10.1007/s11356-021-18419-w

Nowicka, B. (2022). Heavy metal–induced stress in eukaryotic algae—Mechanisms of heavy metal toxicity and tolerance with particular emphasis on oxidative stress in exposed cells and the role of antioxidant response. *Environmental Science and Pollution Research International, 29*(12), 16860–16911. DOI: 10.1007/s11356-021-18419-w

Pant, N., Kumar, G., Upadhyay, A. D., Patel, D. K., Gupta, Y. K., & Chaturvedi, P. K. (2014). Reproductive toxicity of lead, cadmium, and phthalate exposure in men. *Environmental Science and Pollution Research International, 21*(18), 11066–11074. DOI: 10.1007/s11356-014-2986-5

Pardo-Hernández, M., López-Delacalle, M., Martí-Guillen, J. M., Martínez-Lorente, S. E., & Rivero, R. M. (2021). ROS and NO phytomelatonin-induced signaling mechanisms under metal toxicity in plants: A review. *Antioxidants, 10*(5), 775. DOI: 10.3390/antiox10050775

Piotto, F. A., Carvalho, M. E. A., Souza, L. A., Rabêlo, F. H. S., Franco, M. R., Batagin-Piotto, K. D., & Azevedo, R. A. (2018). Estimating tomato tolerance to heavy metal toxicity: Cadmium as study case. *Environmental Science and Pollution Research International, 25*(27), 27535–27544. DOI: 10.1007/s11356-018-2778-4

Popova, L., Maslenkova, L., Yordanova, R., Krantev, A., Szalai, G., & Janda, T. (2008). Salicylic acid protects photosynthesis against cadmium toxicity in pea plants. *General and Applied Plant Physiology, 34*(3-4), 133–148.

Popova, L. P., Maslenkova, L. T., Ivanova, A., & Stoinova, Z. (2011). Role of salicylic acid in alleviating heavy metal stress. In *Environmental adaptations and stress tolerance of plants in the era of climate change* (pp. 447–466). Springer New York.

Rai, G. K., Bhat, B. A., Mushtaq, M., Tariq, L., Rai, P. K., Basu, U., Dar, A. A., Islam, S. T., Dar, T. U., & Bhat, J. A. (2021). Insights into decontamination of soils by phytoremediation: A detailed account on heavy metal toxicity and mitigation strategies. *Physiologia Plantarum, 173*(1), 287–304. DOI: 10.1111/ppl.13433

Rajewska, I., Talarek, M., & Bajguz, A. (2016). Brassinosteroids and response of plants to heavy metals action. *Frontiers in Plant Science*, *7*, 629. DOI: 10.3389/fpls.2016.00629

Rani, A., Kumar, A., Lal, A., & Pant, M. (2014). Cellular mechanisms of cadmium-induced toxicity: A review. *International Journal of Environmental Health Research*, *24*(4), 378–399. DOI: 10.1080/09603123.2013.835032

Rizvi, A., Zaidi, A., Ameen, F., Ahmed, B., AlKahtani, M. D., & Khan, M. S. (2020). Heavy metal induced stress on wheat: Phytotoxicity and microbiological management. *RSC Advances*, *10*(63), 38379–38403. DOI: 10.1039/D0RA05610C

Rucińiska-Sobkowiak, R. (2010). Oxidative stress in plants exposed to heavy metals. *Postępy Biochemii*, *56*(2), 191–200.

Salazar, M., González, E., Casaretto, J. A., Casacuberta, J. M., & Ruiz-Lara, S. (2007). The promoter of the TLC1. 1 retrotransposon from Solanum chilense is activated by multiple stress-related signaling molecules. *Plant Cell Reports*, *26*(10), 1861–1868. DOI: 10.1007/s00299-007-0375-y

Salifu, M., John, M. A., Abubakar, M., Bankole, I. A., Ajayi, N. D., & Amusan, O. (2024). Phytoremediation strategies for heavy metal contamination: A Review on sustainable approach for environmental restoration. *Journal of Environmental Protection*, *15*(4), 450–474. DOI: 10.4236/jep.2024.154026

Schutzendubel, A., & Polle, A. (2002). Plant responses to abiotic stresses: Heavy metal-induced oxidative stress and protection by mycorrhization. *Journal of Experimental Botany*, *53*(372), 1351–1365. DOI: 10.1093/jexbot/53.372.1351

Seneviratne, M., Rajakaruna, N., Rizwan, M., Madawala, H., Ok, Y. S., & Vithanage, M. (2019). Heavy metal-induced oxidative stress on seed germination and seedling development: A critical review. *Environmental Geochemistry and Health*, *41*(4), 1813–1831. DOI: 10.1007/s10653-017-0005-8

Shoaib, A., & Javaid, A. 2021. Oxidative stress in plants exposed to heavy metals. In *Organic Solutes, Oxidative Stress, and Antioxidant Enzymes Under Abiotic Stressors* (pp. 133-152). CRC Press. DOI: 10.1201/9781003022879-7

Shu, X., Yin, L., Zhang, Q., & Wang, W. (2012). Effect of Pb toxicity on leaf growth, antioxidant enzyme activities, and photosynthesis in cuttings and seedlings of Jatropha curcas L. *Environmental Science and Pollution Research International*, *19*(3), 893–902. DOI: 10.1007/s11356-011-0625-y

Shukla, P. K., Misra, P., Maurice, N., & Ramteke, P. W. (2019). Heavy metal toxicity and possible functional aspects of microbial diversity in heavy metal-contaminated sites. In *Microbial Genomics in Sustainable Agroecosystems* (Vol. 2, pp. 255–317). Springer Singapore. DOI: 10.1007/978-981-32-9860-6_15

Simek, J., Tuma, J., Dohnal, V., Musil, K., & Ducaiová, Z. (2016). Salicylic acid and phenolic compounds under cadmium stress in cucumber plants (Cucumis sativus L.). *Acta Physiologiae Plantarum*, *38*(7), 172. DOI: 10.1007/s11738-016-2192-9

Singh, B. N., Mishra, R. N., Agarwal, P. K., Goswami, M., Nair, S., Sopory, S. K., & Reddy, M. K. (2004). A pea chloroplast translation elongation factor that is regulated by abiotic factors. *Biochemical and Biophysical Research Communications*, *320*(2), 523–530. DOI: 10.1016/j.bbrc.2004.05.192

Sofy, M. R., Seleiman, M. F., Alhammad, B. A., Alharbi, B. M., & Mohamed, H. I. (2020). Minimizing adverse effects of pb on maize plants by combined treatment with jasmonic, salicylic acids and proline. *Agronomy (Basel)*, *10*(5), 699. DOI: 10.3390/agronomy10050699

Song, C., Bibi, Z., Ali, S., & Ahmad, R. (2026). Silicon mitigates aluminum toxicity in lettuce by improving photosynthetic pigments and antioxidant defense system. *Biometals : An International Journal on the Role of Metal Ions in Biology, Biochemistry, and Medicine*, •••, 1–14. DOI: 10.1007/s10534-026-00819-5

Song, W. Y., Yang, H. C., Shao, H. B., Zheng, A. Z., & Brestic, M. (2014). The alleviative effects of salicylic acid on the activities of catalase and superoxide dismutase in malting barley (Hordeum uhulgare L.) seedling leaves stressed by heavy metals. *Clean (Weinheim)*, *42*(1), 88–97. DOI: 10.1002/clen.201200310

Sun, L., Wang, R., Ju, Q., Xing, M., Li, R., Li, W., Li, W., Wang, W., Deng, Y., & Xu, J. (2023). Mitigation mechanism of zinc oxide nanoparticles on cadmium toxicity in tomato. *Frontiers in Plant Science*, *14*, 1162372. DOI: 10.3389/fpls.2023.1162372

Sytar, O., Kumar, A., Latowski, D., Kuczynska, P., Strzałka, K., & Prasad, M. N. V. (2013). Heavy metal-induced oxidative damage, defense reactions, and detoxification mechanisms in plants. *Acta Physiologiae Plantarum*, *35*(4), 985–999. DOI: 10.1007/s11738-012-1169-6

Tamás, L., Mistrík, I., Alemayehu, A., Zelinová, V., Bočová, B., & Huttová, J. (2015). Salicylic acid alleviates cadmium-induced stress responses through the inhibition of Cd-induced auxin-mediated reactive oxygen species production in barley root tips. *Journal of Plant Physiology*, *173*, 1–8. DOI: 10.1016/j.jplph.2014.08.018

Tapia, G., Verdugo, I., Yanez, M., Ahumada, I., Theoduloz, C., Cordero, C., Poblete, F., González, E., & Ruiz-Lara, S. (2005). Involvement of ethylene in stress-induced expression of the TLC1. 1 retrotransposon from Lycopersicon chilense Dun. *Plant Physiology*, *138*(4), 2075–2086. DOI: 10.1104/pp.105.059766

Touzout, N., Mehallah, H., Moralent, R., Nemmiche, S., & Benkhelifa, M. (2021). Co-contamination of deltamethrin and cadmium induce oxidative stress in tomato plants (*Solanum lycopersicum* L.). *Acta Physiologiae Plantarum*, *43*(6), 91. DOI: 10.1007/s11738-021-03261-x

Wang, F., Tan, H., Huang, L., Cai, C., Ding, Y., Bao, H., Chen, Z., & Zhu, C. (2021). Application of exogenous salicylic acid reduces Cd toxicity and Cd accumulation in rice. *Ecotoxicology and Environmental Safety*, *207*, 111198. DOI: 10.1016/j.ecoenv.2020.111198

Wani, Z. A., Ahmad, Z., Asgher, M., Bhat, J. A., Sharma, M., Kumar, A., Sharma, V., Kumar, A., Pant, S., Lukatkin, A. S., & Anjum, N. A. (2023). Phytoremediation of potentially toxic elements: Role, status and concerns. *Plants*, *12*(3), 429. DOI: 10.3390/plants12030429

Yadav, S. K. (2010). Heavy metals toxicity in plants: An overview on the role of glutathione and phytochelatins in heavy metal stress tolerance of plants. *South African Journal of Botany*, *76*(2), 167–179. DOI: 10.1016/j.sajb.2009.10.007

Yordanova, R., Maslenkova, L., Paunova, S. and Popova, L., 2009. Sensitivity of photosynthetic apparatus of pea plants to heavy metal stress. *Biotechnology & Biotechnological Equipment, 23*(sup1), pp.347-350.

Zong, H., Liu, J., Wang, F., & Song, N. (2020). Root morphological response of six peanut cultivars to chromium (VI) toxicity. *Environmental Science and Pollution Research International*, *27*(15), 18403–18411. DOI: 10.1007/s11356-020-08188-3

Zulfiqar, A., Gul, B., Saleem, A., Islam, A., Zulfiqar, U., Ali, M. F., Nawaz, M., Al-Ghamdi, A. A., & Rizwana, H. (2024). Salicylic Acid Mitigates Cadmium Stress in Wheat: Experimental Insights Into Growth and Biochemical Parameters. *Scientifica*, *2024*(1), 6887694. DOI: 10.1155/sci5/6887694

Zulfiqar, U., Jiang, W., Xiukang, W., Hussain, S., Ahmad, M., Maqsood, M. F., Ali, N., Ishfaq, M., Kaleem, M., Haider, F. U., Farooq, N., Naveed, M., Kucerik, J., Brtnicky, M., & Mustafa, A. (2022). Cadmium phytotoxicity, tolerance, and advanced remediation approaches in agricultural soils; a comprehensive review. *Frontiers in Plant Science*, *13*, 773815. DOI: 10.3389/fpls.2022.773815

# Chapter 7
# Biochar Potential Against Metal Toxicity in Plants

**Nabil Touzout**
http://orcid.org/0000-0002-7091-9990
*University Yahia Fares of Medea, Medea, Algeria*

**Mohammad Shahid**
*Research Center, Department of Agriculture, Faculty of Science, Marwadi University, Rajkot, India*

**Meriem Babaali**
*Pole Urban Ouzera, University of Medea, Medea, Algeria*

**Aftab Jamal**
http://orcid.org/0000-0001-8518-3130
*The University of Agriculture, Peshawar, Pakistan*

**Hafiza Muniba Din Muhammad**
http://orcid.org/0000-0003-1771-3624
*Bahauddin Zakariya University, Multan, Pakistan*

DOI: 10.4018/979-8-3693-9899-9.ch007

## ABSTRACT

*The stunted growth, poor nutrient uptake, oxidative damage, and reduced physiological functioning are symptoms of metal toxicity in plants. Biochar, a stable carbon-rich substance produced by pyrolyzing organic materials in an oxygen-limited environment, is a potential method for enhancing soil quality under metal stress. Its large surface area, porous structure, and capacity to interact with soil constituents, biochar can improve soil fertility, increase water retention, and lower the bioavailability of harmful metals. Biochar promotes plant growth and developmental processes and increases tolerance to metal toxicity. It explores the fundamental processes by which biochar interacts with soil and plant systems, enhances plant defenses, controls photosynthesis, increases plant growth, and reduces the harmful compounds produced under metal stress. This chapter attempts to validate the potential of biochar as an environmentally sustainable method to reduce the negative effects of metal contamination in agriculture.*

## INTRODUCTION

Heavy metal pollution in soil is a growing worldwide issue that affects food security, environmental sustainability, and agricultural productivity (Khan et al., 2025). Industrial operations, mining operations, inappropriate waste disposal, and the use of polluted water for irrigation have all contributed to the extensive pollution of agricultural soils by toxic metal (Sarim et al., 2022). These contaminants pose serious threats to plant health, growth, and development, in addition to damaging soil ecosystems. (Farooqi et al., 2022). Heavy metal buildup in the soil has a detrimental effect on various plant physiological functions, resulting in stunted growth, poor nutrient uptake, and frequent plant mortality (Hafeez et al., 2022; Sinha et al., 2025; Touzout et al., 2021a). Furthermore, these metals can enter the food chain via plant tissues, posing direct health concerns to animals and humans that consume polluted crops (Zeeshan et al., 2023). Metal toxicity disturbs plant health in several ways. First, they prevent the growth of roots, which limits the plant's ability to absorb water and essential nutrients from the soil (Ahmad et al., 2025). Metals that adhere to cell walls and disrupt enzymatic processes can also cause oxidative stress. Oxidative stress results in the production of reactive oxygen species (ROS), which damage cell lipids, proteins, and DNA, leading to cellular malfunction and plant senescence (Kiran et al., 2021; Touzout et al., 2021b).

Moreover, metal toxicity impairs the growth and development of plants by interfering with their photosynthetic mechanisms, which lowers chlorophyll concentration and overall photosynthetic efficiency (Giannakoula et al., 2021). Because metals do not biodegrade but instead accumulate in ecosystems, their high toxicity and long-term persistence in the environment make the process of cleaning up and rehabilitating polluted soil difficult (Gong et al., 2018). Traditional soil decontamination techniques, such as soil washing, electrokinetic treatment, and phytoremediation, are costly, time-consuming, and might not provide long-term fixes (Jacob et al., 2025). Moreover, using chemicals for clean-up may result in secondary contamination, which contributes to adverse environmental effects (Weldeslassie et al., 2017). Therefore, it is essential to develop sustainable, low-cost, and alternative methods to address metal toxicity in soil. In this regard, biochar, a carbonaceous material produced by pyrolyzing organic waste at high temperatures without oxygen, has been proposed as a potential remedy for enhancing soil health and reducing the harmful effects of metals on plants (Ul Haq et al., 2025). Biochar, a relatively stable, porous substance, has been thoroughly studied for its potential as a soil ameliorant, a source of readily available nutrients, and a way to improve plant growth in challenging environments (Mihoub et al., 2024; Yuan et al., 2019). A large surface area, porous structure, and surface functional groups are some of the special physical and chemical characteristics of biochar that allow it to interact with soil constituents, adsorb harmful compounds, and lower the bioavailability of heavy metals (Ahmad et al., 2022). Therefore, biochar is a useful technique for reducing metal toxicity in plants and provides a sustainable and affordable way to increase agricultural output in contaminated soils (Bayar et al., 2024).

Biochar helps plant growth by enhancing the structure of the soil, retaining more water, and ensuring that nutrients are released (Ul Haq et al., 2025). Furthermore, it has been found that biochar reduces the absorption of heavy metals in plants, thereby inhibiting their toxic activities. By adsorbing the metals on its surface and immobilizing them in the soil, where they cannot be taken up by root systems, biochar reduces the oxidative stress caused by heavy metals and promotes healthy plant growth (Imran et al., 2025). Because it can increase a plant's resistance to other stressors such as drought, salinity, and temperature, biochar is a multipurpose option for agricultural tolerance against climate change (Touzout et al., 2025; Zulfiqar et al., 2022). Although biochar has long been used in agriculture, current research has shown that it may help reduce plant metal toxicity. Numerous studies have shown that biochar increases photosynthetic activity, encourages root growth, lowers the concentration of harmful metals in the rhizosphere, and increases the total biomass of plants (Anbuganesan et al., 2024; R. Wu et al., 2025). In addition to reducing metal toxicity, biochar improves nutrient uptake, soil microbial activity, and activation of antioxidant defense mechanisms in plants (Alasmary et al., 2025). Biochar is, therefore, a flexible solution for metal toxicity in agricultural systems (Luo et al., 2025).

The goal of this chapter is to present a thorough analysis of recent research on biochar's potential to reduce metal toxicity in plants. The basic processes by which biochar interacts with soil and plants, its effects on plant growth and development, and its capacity to increase plant resistance to environmental stressors are fully explored in this chapter. Moreover, it will cover the different ways that biochar can be used in soils that are affected by metals, the way it can increase agricultural output, and where it fits into sustainable farming methods. Farmers can more effectively use biochar as farming practices to address the problem of metal pollution and improve sustainable production

## SIGNIFICANCE OF BIOCHAR IN PLANT GROWTH AND DEVELOPMENTAL PROCESSES

Biochar has gained attention as a green multipurpose amendment with a significant impact on plant growth and development. Because of its wide surface area, porous shape, and stable carbon content, biochar, which is produced by burning organic materials under oxygen-limited or oxygen-free conditions, offers various agronomic advantages (Kuryntseva et al., 2023). An ideal environment for seed germination, root development, nutrient absorption, and plant stability under metal stress is developed when biochar is added to agricultural soils (Anbuganesan et al., 2024). This improves both the physical and chemical properties of the soil matrix (Ren et al., 2025). In terms of physical properties, the porous structure of biochar improves soil aeration and water-holding capacity (Jia et al., 2024). This is especially helpful in clayey soils, where compaction can inhibit root growth, and in sandy soils, where water drainage is excessive (Blanco-Canqui, 2017). Biochar increases the soil's ability to retain water, which makes moisture available at the root zone, a critical factor during drought and seedling establishment (Gul-Lalay et al., 2024). Furthermore, improved root penetration, growth, and retention are made possible by enhanced soil porosity caused by the use of biochar. These factors are critical for nutrient exploration and plant stabilization (Alkharabsheh et al., 2021). Biochar contributes to the soil nutrient dynamics. One of its important roles is to increase the soil's cation exchange capacity (CEC), which enables it to store essential minerals such as potassium ($K^+$), calcium ($Ca^{2+}$), magnesium ($Mg^{2+}$), and ammonium ($NH_4^+$) and release them gradually when the plant needs them (Ul Haq et al., 2025). Its nutrient buffer capability reduces the possibility of nutrient leakage, especially in areas with high rainfall or light-textured soils (Adnyana et al., 2025). Furthermore, biochar can change the pH of acidic soils; it usually raises the pH, increasing the availability of phosphorus and other micronutrients (Diaz-Chuquizuta et al., 2025).

To maximize nutrient solubilization and enzymatic activity in the rhizosphere, pH adjustment is essential. Functional groups attached to the biochar surface also aid in nutrient adsorption and desorption, thereby increasing the effectiveness of nutrient usage (Van Nguyen et al., 2025). Biochar has been shown to function as a substrate for beneficial soil microbes at the biological level. Microbial communities such as nitrogen-fixing bacteria, mycorrhizal fungi, and plant growth-promoting rhizobacteria (PGPR) reduce metal toxicity (Hafeez et al., 2022). These microbes are essential for the generation of phytohormones, phosphate solubilization, and nitrogen fixation, all of which directly impact the growth and development of plants (Gul-Lalay et al., 2024). Utilizing biochar for microbial activation also stimulates soil enzymatic activities, such as urease, phosphatase, and dehydrogenase, which are crucial for organic matter decomposition and nutrient turnover. A more dynamic and fertile soil environment with higher plant productivity supports such improved biological processes (Palansooriya et al., 2019).

Biochar can be engineered to modify a variety of internal processes in plant physiology and development. It has been shown that biochar alters the expression of genes related to photosynthesis, stress response pathways, and nutrient transporters (Xu et al., 2024). Higher levels of carotenoids, chlorophyll, and other photosynthetic pigments have also been linked to biochar treatments, suggesting higher photosynthetic activity. Increased biomass accumulation, leaf area, and yield potential are typically the result of improved physiological growth and nutrition. Treated plants absorb more water and nutrients because of their increased root surface area and root hair development (Antonangelo et al., 2025). Additionally, biochar has been utilized to combat abiotic stressors, including salt, drought, and toxicities caused by heavy metal pollution (Gul-Lalay et al., 2024; Mihoub et al., 2024). Because of its superior adsorption capabilities, it can absorb harmful substances from the soil and limit their availability to plants. Furthermore, it has been discovered that biochar activates antioxidant enzymes, which lessen oxidative damage caused by oxidative stress. Generally, in addition to improving plant health, an increase in antioxidants supports cellular integrity under metal toxicity (Touzout et al., 2025).

Biochar affects the vegetative and reproductive stages of plant growth from a developmental perspective. Dense vegetative growth is caused by excess nutrients and water, whereas improved physiological status promotes flowering and fruit set. Additionally, it has been demonstrated that biochar affects the synthesis of phytohormones that control processes such as cell division, elongation, and differentiation, including auxins, gibberellins, and cytokinins (Mushtaq et al., 2025). High-yielding crops require a more uniform growth habit, more branching, and perhaps more synchronized blooming, all of which are induced by these hormonal effects. The application of biochar in agricultural systems is a very promising method for improving plant development and growth by combining biological, chemical, and physical processes. Its function as a multipurpose modulator of the plant-soil-microbe interface extends beyond that of a soil conditioner. Biochar is a promising way to boost crop resilience, optimize resource usage, and maintain production in an era of growing climate change and soil erosion-related agricultural issues (Figure 1). To fully utilize biochar in contemporary agroecosystems, further research on its interactions with various soils, crops, and environmental factors is necessary.

*Figure 1. Schematic diagram illustrates the biochar-mediated mitigation of metal toxicity in plants*

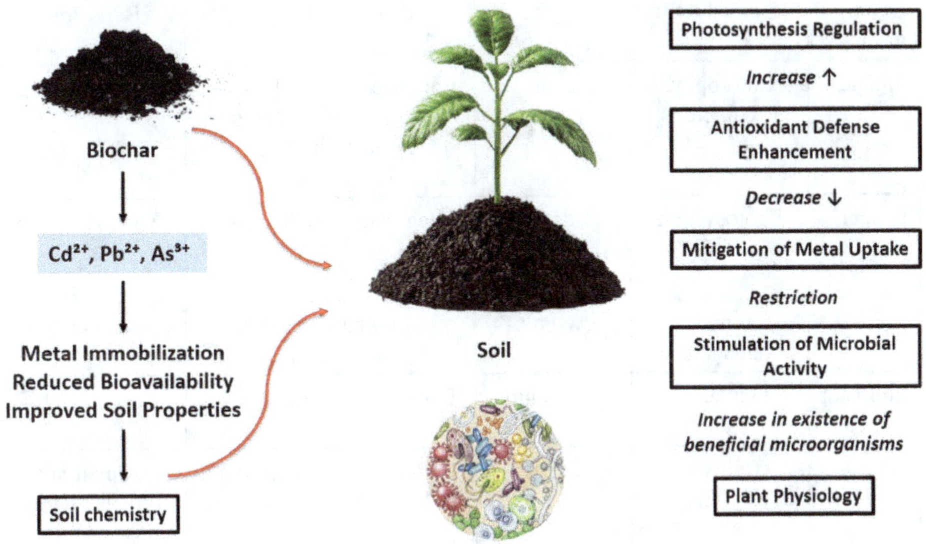

## BIOCHAR REGULATES PHOTOSYNTHETIC MACHINERY IN PLANTS SUBJECTED TO METAL TOXICITY

*Table 1. Potential of biochar in the regulation of photosynthesis system under metal toxicity in plants*

| Crop names | Affected traits | Metal toxicity | Biochar impacts | References |
|---|---|---|---|---|
| Spinach | Chlorophyl content | Lead | Enhances chlorophyll and carotenoid contents Protection of cell membrane | Iqbal et al. (2024) |
| Pepper | Photosystem activity | Cadmium | Regulates efficiency of photosystem II in stressed plants | Xu et al. (2016) |
| Chilli | Stomatal conductance | Arsenic | Maintains stomatal conductance | Manzoor et al. (2024) |
| Eggplant | Oxidative stress | Cadmium | Boost scavengers of toxic elements | Li et al. (2016) |
| Wheat | Rubisco activity | Arsenic | Reduce metal uptake and boost enzymatic function | Kumar and Bhattacharya (2022) |
| Maize | Chloroplast ultrastructure | Zinc, copper, and nickel | Protection of thylakoid membranes and grana stacking | Boros-Lajszner et al. (2025) |

The photosynthetic function is the primary physiological activity that drives growth and biomass accumulation, and heavy metal contamination in agricultural soils poses a serious threat to plant productivity (Table 1). Through a variety of mechanisms, including oxidative pigment breakdown, photosystem II activity, destruction of the ultrastructure of chloroplasts, and impairment of electron transport, metals interfere with the photosynthetic apparatus (Jha, 2023; Touzout et al., 2024a). Accordingly, biochar has become a viable strategy for protecting and improving the photosynthetic machinery of plants under metal stress, in addition to immobilizing harmful metals in the rhizosphere (Zhao et al., 2025). Biochar's regulatory function starts at the soil-soil interface, where it modifies the metal availability and chemical dynamics. Using surface complexation, electrostatic attraction,

ion exchange, and precipitation, biochar binds heavy metals because of its large surface area, functional groups (such as carboxyl, hydroxyl, and phenolic groups), and alkaline nature. This lessens the bioavailability of harmful ions in the root zone, which minimizes their transfer to the aerial regions where photosynthesis takes place (Luo et al., 2025). To mitigate their impact on chlorophyll synthesis and light harvesting efficiency, biochar can lower shoot concentrations of Cd and Pb by more than 50% (Angon et al., 2024).

Biochar use in metal-stressed plants may have been linked to notable improvements in photosynthetic metrics at the physiological level. These include improved water-use efficiency and higher levels of intercellular $CO_2$ concentration (Ci), transpiration rate (E), stomatal conductance (gs), and net photosynthetic rate (Pn) (Chen et al., 2023). Increased levels of carotenoids, which are normally broken down under metal stress, and chlorophyll a and b are frequently observed in conjunction with these effects. Biochar contributes to the maintenance of light absorption capacity and photochemical efficiency by maintaining pigment integrity, both of which are essential for electron transport and ATP synthesis (Libutti et al., 2020). Furthermore, biochar reduces oxidative damage to chloroplasts caused by ROS, which is a frequent side effect of metal toxicity. Under metal stress, the overproduction of ROS leads to lipid peroxidation, protein denaturation, and damage to chloroplast membranes. By strengthening the antioxidant defense mechanism of the plant, biochar indirectly lessens these effects. It increases the activity of antioxidant enzymes that scavenge ROS and safeguard the photosynthetic machinery and antioxidants. Biochar-treated plants under metal stress have shown antioxidant enzyme activities comparable to or higher than those of unstressed control plants (Abideen et al., 2020; Zulfiqar et al., 2022).

Biochar contributes to osmotic adjustment and cellular homeostasis through the modulation of compatible solutes and enzymatic defense systems, which also play protective roles for photosynthetic proteins and membranes. Furthermore, biochar may influence the expression of genes involved in photosynthesis-related pathways. This molecular-level regulation reflects the broader capacity of biochar to influence gene expression and metabolic adaptation under unfavorable conditions (Hareem et al., 2024; Xu et al., 2024).

The impact of biochar on nutrient dynamics is another way it controls photosynthesis under metal toxicity. Deficits in magnesium (Mg), iron (Fe), and phosphorus (P), which are necessary for the transmission of energy and chlorophyll production, are common in metal-stressed plants (Adnyana et al., 2025). Through enhanced CEC and pH buffering, biochar enhances nutrient availability and retention, establishing a more nutrient-friendly environment for plants, even when heavy metals are present. For instance, biochar has been demonstrated to restore Mg and Fe uptake in Cd-contaminated soils, sustain chloroplast activity, and increase chlorophyll content (Razzaq et al., 2022). From a structural perspective, biochar may help to maintain chloroplast integrity when exposed to metal stress. Plants exposed to heavy metals without biochar frequently display deformed chloroplasts, damaged thylakoid membranes, and accumulation of plastoglobuli, which are signs of stress-induced damage, according to transmission electron microscopy (TEM) investigations (Rana et al., 2025). On the other hand, plants treated with biochar retain more consistent chloroplast morphology, with membranes that remain intact and well-organized grana, which is associated with improved photosynthetic efficiency.

Biochar can improve rhizospheric microbial communities that aid in metal detoxification and nutrient cycling, in addition to its direct impact on plant physiology. When supported by charcoal amendment, specific plant growth-promoting rhizobacteria (PGPR) and arbuscular mycorrhizal fungi (AMF) can help decrease metal uptake and enhance nutrient assimilation, thereby indirectly maintaining photosynthetic activity. This three-way relationship between microorganisms, plants, and biochar is an integrated approach for managing metal stress (Palansooriya et al., 2019; Hafeez et al., 2022). The photosynthetic machinery of plants exposed to metal toxicity is regulated by biochar through a variety of mechanisms, such as metal immobilization, antioxidant enhancement, nutritional balancing, gene expression control, and chloroplast structural protection. It is a useful tool in phytoremediation techniques and sustainable agriculture because of its capacity to reduce the detrimental effects of heavy metals on photosynthesis, especially in regions where soil contamination is a problem. Its use for safeguarding and improving photosynthetic performance under environmental stress will be further refined by future studies that concentrate on plant-specific responses, biochar types, and interactions with microbial communities.

## BIOCHAR CAN LESSEN THE TOXIC SUBSTANCES GENERATED UNDER METAL TOXICITY IN PLANTS

A prevalent environmental problem is heavy metal poisoning, particularly in farming areas where mining, industrial operations, wastewater irrigation, and the overuse of chemical fertilizers have an impact. When plants are exposed to high levels of heavy metals such as arsenic (As), cadmium (Cd), lead (Pb), and mercury (Hg), they undergo significant physiological and biochemical stress, which frequently leads to an excess of poisonous chemicals. These reactive oxygen species (ROS) can harm cellular constituents, such as membranes, proteins, lipids, and nucleic acids. ROS includes hydrogen peroxide ($H_2O_2$), superoxide anions ($O_2^-$), and hydroxyl radicals (•OH). Additionally, metal stress can throw off metabolic equilibrium and cause the production of harmful metabolites, such as phytotoxic organic acids or methylated metal compounds (Demidchik, 2015; Rathinapriya et al., 2024; Touzout et al., 2024b). Biochar, a porous, carbon-rich material made from pyrolyzed biomass, has been the focus of extensive research in recent years due to its potential to reduce the toxicity of heavy metals to plants. In this regard, reducing the production and buildup of harmful compounds in plant tissues is one of the most crucial roles of biochar. This is accomplished by combining biological, physical, and chemical mechanisms that function both inside the plant system and at the soil-plant interface (Imran et al., 2025). Biochar effectively lowers the rhizosphere concentration of harmful metals (Wu et al., 2025). Through sorption, ion exchange, complexation, and precipitation processes, it can interact well with metal ions due to its wide surface area, negative charge, and high density of oxygen-containing functional groups (hydroxyl, carboxyl, and carbonyl). As a result, there is a lower free metal ion concentration in the soil solution, which lowers the plant root uptake. In plant cells, this reduced internal metal load aids prevent the emergence of metal-induced oxidative stress and the subsequent production of toxic ROS (Godwin et al., 2019). Second, biochar strengthens the antioxidant defense system of plants, which is an essential biochemical mechanism to detoxify harmful compounds produced by metal stress. The charcoal mixed with the growing medium increased the activity of antioxidant enzymes.

Together, these enzymes detoxify superoxide radicals into less dangerous compounds, which are subsequently converted into molecular oxygen and water. The amount of non-enzymatic antioxidants, which chelate heavy metals and directly combat free radicals, can also be increased by biochar (Ghazouani et al., 2023; Zulfiqar et al., 2022). The regulation of secondary metabolite pathways is another primary function of biochar. Plants that are under metal stress build up secondary metabolites, such as phenolics, alkaloids, and lignin derivatives, which are poisonous in large concentrations and inhibit regular metabolic functions. By suppressing stress signaling and metal uptake, biochar indirectly modifies these pathways, enabling the generation of more balanced secondary metabolites. It permits improved plant development and lowers the metabolic cost of detoxification procedures (He et al., 2024). Metal speciation in plant tissues is also affected by biochar. The addition of biochar has occasionally been linked to the transformation of extremely reactive metal ions into less harmful or inert forms. To prevent the disruption of cellular metabolism, biochar can promote metal-binding peptides that sequester heavy metals in the vacuole or cell wall (Imran et al., 2025). In addition to slowing down the production of ROS, this sequestration response prevents the production of additional harmful byproducts, such as lipid peroxidation end products, malondialdehyde. Furthermore, reducing the harmful consequences of metal toxicity requires an understanding of the impact of biochar on plant nutritional status.

Nutrient imbalances, such as those involving deficits in vital elements like calcium (Ca), magnesium (Mg), potassium (K), and iron (Fe), which are important for maintaining membrane stability, enzymatic activities, and chlorophyll production, are likely to be brought on by metal toxicity. It has been discovered that adding biochar to soils improves the availability and assimilation of these nutrients, protecting plant cells against oxidative damage and the accumulation of harmful metabolites (Wu et al., 2023). In addition to directly affecting plant physiology, biochar promotes advantageous alterations in the rhizosphere microbial community. It increases the growth of mycorrhizal fungi and plant growth-promoting rhizobacteria (PGPR), the majority of which play a part in lessening metal stress. Siderophores, organic acids, and biosurfactants that chelate heavy metals or change their mobility may be produced by microbes. Some also cause plants to develop systemic resistance, which improves their ability to withstand hazardous substances (Hafeez et al., 2022). Finally, the buffering ability of biochar restores soil pH, which is otherwise disturbed by metal stress. Soil acidification tends to increase the toxicity and solubility of metals, which in turn increases the production of hazardous metabolites in plants. Biochar limits metal solubility, inhibits root uptake, and stops stress reactions that could result in the development of hazardous byproducts by maintaining the pH between neutral and slightly alkaline (Antonangelo et al., 2025). Biochar has a multifaceted capacity to reduce harmful compounds produced by heavy metal toxicity in plants. It works in the following ways: (1) decreasing the uptake and translocation of metals, (2) strengthening antioxidant defenses, (3) regulating the production of metabolites linked to stress, (4) stabilizing metal species in plant tissues, (5) boosting nutrient uptake, and (6) promoting beneficial soil microbes. Together, these benefits make biochar an excellent supplement for sustainable agriculture and phytoremediation, especially in contaminated soils. For optimal detoxification effectiveness, future studies must focus on optimizing the type of biochar and application rates for the target metal contaminants and plant species (Table 2).

*Table 2. Role of biochar in the reduction of toxic substances produced under metal toxicity in plants*

| Crop name | Toxic elements | Metal toxicity | Biochar impact | References |
|---|---|---|---|---|
| Rice | Reactive oxygen species (ROS) | Nickel | Elevation of antioxidant enzymes | Ali et al. (2020) |
| Rice | Hydrogen Peroxide ($H_2O_2$) | Cadmium | Diminishes oxidative injury and continues redox balance | Ali et al. (2024) |
| Sorghum | Malondialdehyde (MDA) | Cadmium | Lesser lipid peroxidation and protection of membrane integrity | Hassan et al. (2020) |

continued on following page

*Table 2. Continued*

| Crop name | Toxic elements | Metal toxicity | Biochar impact | References |
|---|---|---|---|---|
| Chilli | Electrolyte leakage | Arsenic | Restore cellular functions | Manzoor et al. (2024) |
| Pepper | Free metal ions | Cadmium | Improves adsorption and immobilization level | Xu et al. (2016) |

## BIOCHAR BOOSTS THE PLANT DEFENSE SYSTEM AGAINST METAL TOXICITY

One of the biggest issues facing modern agriculture is plant heavy metal stress, which is particularly problematic for mining, industrial effluent pollution, wastewater reuse, and excessive pesticide use. When plants are exposed to dangerous metals like lead (Pb), arsenic (As), chromium (Cr), mercury (Hg), or cadmium (Cd), their cellular equilibrium is severely disrupted (Rubiya et al., 2018). Overproduction of reactive oxygen species (ROS), which results in oxidative stress and damage to cellular components such as lipids, proteins, and nucleic acids, is one of the main effects. In response to this type of stress, plants initiate a multilayered defensive system that includes metal chelators, stress proteins, signaling pathways, and enzymatic and non-enzymatic antioxidants. Plant defense mechanisms against metal stressors can be strengthened by biochar, a carbonaceous solid derived from pyrolyzed biomass (Ghazouani et al., 2023). Reducing the availability of harmful metal ions in the rhizosphere is one of the most significant effects of biochar. Particles of biochar have a large surface area and contain a variety of functional groups, including phenolic, carboxyl, and hydroxyl groups. Through mechanisms such as adsorption, ion exchange, and complexation, these functions enable biochar to adsorb metal ions, thereby decreasing their mobility and bioavailability. Biochar indirectly aids in plant defense by lowering the internal metal load and the level of cellular damage caused by limiting the amount of metal ions absorbed by plant roots (Godwin et al., 2019).

In addition to immobilization, biochar actively strengthens the antioxidant systems of plants. The basis of plants' ability to withstand metal stress is their antioxidant defense mechanisms. It contains essential enzymes. Together, these enzymes detoxify ROS and protect cells from oxidative damage. Studies have demonstrated that treating contaminated soil with biochar causes these enzymes to be widely upregulated in crops. For example, crops cultivated in soil contaminated with cadmium and supplemented with biochar exhibit increased antioxidant activity, together with a reduction in lipid peroxidation and hydrogen peroxide accumulation (Zulfiqar et al., 2022). Biochar increases the amount of non-enzymatic protection chemicals and enzymatic antioxidants. They consist of flavonoids, phenolic acids, glutathione (GSH), and ascorbic acid (vitamin C). These molecules serve two purposes: they are metal chelators that keep harmful ions out of delicate cellular locations, and they efficiently scavenge free radicals (Touzout et al., 2024c). Its improvement by adding biochar not only increases resistance to oxidative stress but also fortifies other defense metabolic processes, including the synthesis of secondary metabolites and lignin, which are involved in stress signaling and structural support (He et al., 2024).

Modification of stress signaling pathways is another important way that biochar shapes plant defenses. Heavy metal exposure triggers a series of signals in plants, including $Ca^{2+}$ ions, NO, $H_2O_2$, and stress hormones such as abscisic acid (ABA), salicylic acid (SA), and jasmonic acid (JA). Biochar can alter these signaling cascades by either boosting positive reactions or buffering excessive signals. For instance, secondary messengers of stress that are toxic in excess can be modulated by biochar (Mushtaq et al., 2025). Furthermore, biochar had a positive impact on the expression of genes related to defense and detoxification. Biochar increases the expression of genes that code for metal-binding proteins, antioxidant enzymes, and metal vacuolar sequestration transporters (Garg et al., 2025). These chemical processes lessen the harmful effects of metals on the cytosol by sequestering them into more inert cellular sites, such as cell walls or vacuoles (Xu et al., 2024). Increased nutrient availability and soil microbial health are additional chemical and physical advantages of biochar that indirectly support plant defenses. Essential nutrients, including phosphorus, potassium, calcium, magnesium, and iron, all of which are typically antagonistic to harmful metals, can be made more readily available by biochar.

The production of defensive proteins and enzymes is enhanced in healthy plants. Furthermore, biochar stimulates the growth of helpful microorganisms in soils, such as mycorrhizal fungi and nitrogen-fixing bacteria, which are advantageous for plant growth and systemic defense. According to previous reports, certain microorganisms in soils treated with biochar release organic acids and siderophores that chelate heavy metals and reduce their toxicity to plants (Palansooriya et al., 2019). By integrating physical, chemical, biological, and molecular processes, biochar strengthens plant defenses against metal toxicity. Reduced metal uptake, antioxidant responses, improved non-enzymatic detoxifiers, stress signal regulation, and initiation of metal tolerance genes are all effects of biochar. Meanwhile, biochar improves plant tolerance to stress by facilitating symbiotic microbial partnerships and increasing nutritional availability. Therefore, biochar is an excellent additive for environmentally friendly farming practices and phytoremediation techniques in soils affected by metals.

## CONCLUSION

Metal toxicity reduces global crop productivity. Polluted soils are a major threat to food production and safety. Stunted growth and irregular developmental processes occur due to metal toxicity in plants. Mitigation of the adverse effects of metal toxicity requires time to achieve higher crop productivity. To feed a huge population in recent times, mainly based on food production and the safety of toxic elements. Biochar is an effective way to reduce the adverse effects of metal toxicity in plants. Farmers can obtain higher production with biochar supplementation. Biochar has excellent capability to absorb metals as well as restrict their translocation in other plant sparts.

## REFERENCES


Abideen, Z., Koyro, H. W., Huchzermeyer, B., Ansari, R., Zulfiqar, F., & Gul, B. (2020). Ameliorating effects of biochar on photosynthetic efficiency and antioxidant defence of Phragmites karka under drought stress. *Plant Biology*, *22*(2), 259–266. DOI: 10.1111/plb.13054

Adnyana, I. M., Bimantara, P. O., & Roni, N. G. K. (2025). Enhancing Soil Fertility Through Azolla Incorporation: Impacts on Nitrogen Cycling and Cation Exchange Capacity. *Organic Farming*, *11*(1), 1–12. DOI: 10.56578/of110101

Ahmad, I., Ghaffar, A., Zakir, A., Khan, Z. U. H., Saeed, M. F., Rasool, A., Jamal, A., Mihoub, A., Marzeddu, S., & Boni, M. R. (2022). Activated Biochar Is an Effective Technique for Arsenic Removal from Contaminated Drinking Water in Pakistan. *Sustainability (Basel)*, *14*(21), 14523. DOI: 10.3390/su142114523

Ahmad, I., Mehmood, S., Muneer, M. A., Jamal, A., Farooq, A., Saeed, M. F., Ali, Q., Daraz, U., Touzout, N., Mihoub, A., & Zhou, J. (2025). Bacillus Inoculation Along with Biogas Slurry Mitigates Nickel Stress in Raphanus Sativus by Regulating Ni Accumulation and the Antioxidant Defense System. *Water, Air, and Soil Pollution*, *236*(11), 715. DOI: 10.1007/s11270-025-08307-1

Alasmary, Z., Majrashi, M. A., Aziz, M. A., & Mumtaz, B. (2025). Combined application of biochar and PGPR alleviates Cd stress in wheat by improving antioxidant defense mechanism and crop physiology. *International Journal of Phytoremediation*, *27*(14), 1–16. DOI: 10.1080/15226514.2025.2531260

Alkharabsheh, H. M., Seleiman, M. F., Battaglia, M. L., Shami, A., Jalal, R. S., Alhammad, B. A., Almutairi, K. F., & Al-Saif, A. M. (2021). Biochar and its broad impacts in soil quality and fertility, nutrient leaching and crop productivity: A review. *Agronomy (Basel)*, *11*(5), 993. Advance online publication. DOI: 10.3390/agronomy11050993

Anbuganesan, V., Vishnupradeep, R., Bruno, L. B., Sharmila, K., Freitas, H., & Rajkumar, M. (2024). Combined Application of Biochar and Plant Growth-Promoting Rhizobacteria Improves Heavy Metal and Drought Stress Tolerance in Zea mays. *Plants*, *13*(8), 1143. DOI: 10.3390/plants13081143

Angon, P. B., Islam, M. S., Kc, S., Das, A., Anjum, N., Poudel, A., & Suchi, S. A. (2024). Sources, effects and present perspectives of heavy metals contamination: Soil, plants and human food chain. *Heliyon*, *10*(7), e28357. DOI: 10.1016/j.heliyon.2024.e28357

Antonangelo, J. A., Sun, X., & Eufrade-Junior, H. de J. (2025). Biochar impact on soil health and tree-based crops: A review. *Biochar*, *7*(1), 51. DOI: 10.1007/s42773-025-00450-6

Bayar, J., Ali, N., Dong, Y., Ahmad, U., Anjum, M. M., Khan, G. R., Zaib, M., Jalal, A., Ali, R., & Ali, L. (2024). Biochar-based adsorption for heavy metal removal in water: A sustainable and cost-effective approach. *Environmental Geochemistry and Health*, *46*(11), 428. DOI: 10.1007/s10653-024-02214-w

Blanco-Canqui, H. (2017). Biochar and Soil Physical Properties. *Soil Science Society of America Journal*, *81*(4), 687–711. DOI: 10.2136/sssaj2017.01.0017

Chen, R., Zheng, L., Zhao, J., Ma, J., & Li, X. (2023). Biochar Application Maintains Photosynthesis of Cabbage by Regulating Stomatal Parameters in Salt-Stressed Soil. *Sustainability (Basel)*, *15*(5), 4206. Advance online publication. DOI: 10.3390/su15054206

Demidchik, V. (2015). Mechanisms of oxidative stress in plants: From classical chemistry to cell biology. *Environmental and Experimental Botany*, *109*, 212–228. DOI: 10.1016/j.envexpbot.2014.06.021

Diaz-Chuquizuta, H., Coral-Cardenas, S. A., Arévalo-Aranda, Y. G., Sánchez-Ojanasta, M., Diaz-Chuquizuta, P., Ocaña-Reyes, J. A., Solórzano Acosta, R. A., & Cuevas Giménez, J. P. (2025). Inoculation of Bacillus subtilis in acidic soil amended with biochar and liming materials in maize cultivation. *Scientific Reports*, *15*(1), 28678. DOI: 10.1038/s41598-025-13230-w

Farooqi, Z. U. R., Ahmad, I., Abdul Qadir, A., Murtaza, G., Rafiq, S., Jamal, A., Zeeshan, N., Murtaza, B., Javed, W., Radicetti, E., & Mancinelli, R. (2022). Zeolite-Assisted Immobilization and Health Risks of Potentially Toxic Elements in Wastewater-Irrigated Soil under Brinjal (Solanum melongena) Cultivation. *Agronomy (Basel)*, *12*(10), 2433. DOI: 10.3390/agronomy12102433

Garg, A., Chauhan, P., Kaur, C., Arora, P. K., Garg, S. K., Singh, V. P., Singh, K. P., & Srivastava, A. (2025). Comprehensive heavy metal remediation mechanisms with insights into CRISPR-Cas9 and biochar innovations. *Biodegradation*, *36*(4), 69. DOI: 10.1007/s10532-025-10165-x

Ghazouani, H., Ibrahimi, K., Amami, R., Helaoui, S., Boughattas, I., Kanzari, S., Milham, P., Ansar, S., & Sher, F. (2023). Integrative effect of activated biochar to reduce water stress impact and enhance antioxidant capacity in crops. *The Science of the Total Environment*, *905*(September), 166950. DOI: 10.1016/j.scitotenv.2023.166950

Giannakoula, A., Therios, I., & Chatzissavvidis, C. (2021). Effect of lead and copper on photosynthetic apparatus in citrus (Citrus aurantium l.) plants. the role of antioxidants in oxidative damage as a response to heavy metal stress. *Plants*, *10*(1), 1–14. DOI: 10.3390/plants10010155

Godwin, P. M., Pan, Y., Xiao, H., & Afzal, M. T. (2019). Progress in Preparation and Application of Modified Biochar for Improving Heavy Metal Ion Removal From Wastewater. *Journal of Bioresources and Bioproducts*, *4*(1), 31–42. DOI: 10.21967/jbb.v4i1.180

Gong, Y., Zhao, D., & Wang, Q. (2018). An overview of field-scale studies on remediation of soil contaminated with heavy metals and metalloids: Technical progress over the last decade. *Water Research*, *147*, 440–460. DOI: 10.1016/j.watres.2018.10.024

Gul-Lalay, , Ullah, S., Shah, S., Jamal, A., Saeed, M. F., Mihoub, A., Zia, A., Ahmed, I., Seleiman, M. F., Mancinelli, R., & Radicetti, E. (2024). Combined Effect of Biochar and Plant Growth-Promoting Rhizbacteria on Physiological Responses of Canola (Brassica napus L.) Subjected to Drought Stress. *Journal of Plant Growth Regulation*, *43*(6), 1814–1832. DOI: 10.1007/s00344-023-11219-1

Hafeez, A., Rasheed, R., Ashraf, M. A., Qureshi, F. F., Hussain, I., & Iqbal, M. (2022). Effect of heavy metals on growth, physiological and biochemical responses of plants. In *Plants and their Interaction to Environmental Pollution: Damage Detection, Adaptation, Tolerance, Physiological and Molecular Responses* (pp. 139–159). Elsevier. DOI: 10.1016/B978-0-323-99978-6.00006-6

Hareem, M., Danish, S., Pervez, M., Irshad, U., Fahad, S., Dawar, K., Alharbi, S. A., Ansari, M. J., & Datta, R. (2024). Optimizing chili production in drought stress: Combining Zn-quantum dot biochar and proline for improved growth and yield. *Scientific Reports*, *14*(1), 1–15. DOI: 10.1038/s41598-024-57204-w

He, T., Chen, L., Wu, Y., Wang, J., Wu, Q., Sun, J., Ding, C., Zhou, T., Chen, L., Jin, A., Li, Y., & Zhu, Q. (2024). Combined Metabolome and Transcriptome Analyses of Maize Leaves Reveal Global Effect of Biochar on Mechanisms Involved in Anti-Herbivory to Spodoptera frugiperda. *Metabolites*, *14*(9), 498. Advance online publication. DOI: 10.3390/metabo14090498

Imran, S., Sumi, M. J., Harine, I. J., Paul, N. C., Mahamud, M. A., Rabbi, R. H. M., Brestic, M., & Rhaman, M. S. (2025). Biochar Amendments in Soil: A Sustainable Approach for Mitigating Heavy Metal Stress in Plants. *Phyton*, *94*(4), 1073–1109. DOI: 10.32604/phyton.2025.064046

Jacob, D. E., Jacob, I. D., Daniel, K. S., Gideon, I. K., & Akpan, U. U. (2025). Innovations in Soil Remediation and Land Restoration. In *Environmental Science and Engineering: Vol. Part F513* (pp. 259–313). Springer. DOI: 10.1007/978-3-031-85327-2_9

Jha, Y. (2023). Regulation of photosynthesis under stress. In *Improving Stress Resilience in Plants: Physiological and Biochemical Basis and Utilization in Breeding* (pp. 35–48). Elsevier. DOI: 10.1016/B978-0-443-18927-2.00002-9

Jia, A., Song, X., Li, S., Liu, Z., Liu, X., Han, Z., Gao, H., Gao, Q., Zha, Y., Liu, Y., Wu, X., & Wang, G. (2024). Biochar enhances soil hydrological function by improving the pore structure of saline soil. *Agricultural Water Management*, *306*, 109170. DOI: 10.1016/j.agwat.2024.109170

Khan, A. R., Mihoub, A., Jamal, A., Saeed, M. F., Soltani-Gerdefaramarzi, S., & Ahamad, M. I. (2025). Assessing Heavy Metal Contamination and Health Risks in Urban Street Dust: Insights from Pakistan's Industrial Growth. *Water, Air, and Soil Pollution*, *236*(11), 738. DOI: 10.1007/s11270-025-08359-3

Kiran, B., Bharti, R., & Sharma, R. (2021). Effect of heavy metals: An overview. *Materials Today: Proceedings*, *51*, 880–885. DOI: 10.1016/j.matpr.2021.06.278

Kuryntseva, P., Karamova, K., Galitskaya, P., Selivanovskaya, S., & Evtugyn, G. (2023). Biochar Functions in Soil Depending on Feedstock and Pyrolyzation Properties with Particular Emphasis on Biological Properties. *Agriculture*, *13*(10), 2003. Advance online publication. DOI: 10.3390/agriculture13102003

Libutti, A., Trotta, V., & Rivelli, A. R. (2020). Biochar, vermicompost, and compost as soil organic amendments: Influence on Growth Parameters, Nitrate and Chlorophyll Content of Swiss Chard (Beta vulgaris L. var. cycla). *Agronomy (Basel)*, *10*(3), 346. DOI: 10.3390/agronomy10030346

Luo, P., Zhang, W., Xiao, D., Hu, J., Li, N., & Yang, J. (2025). Biochar-Based Fertilizers: Advancements, Applications, and Future Directions in Sustainable Agriculture—A Review. *Agronomy (Basel)*, *15*(5), 1104. DOI: 10.3390/agronomy15051104

Mihoub, A., Mesnoua, M., Touzout, N., Zeguerrou, R., Siabdallah, N., Benchikh, C., Benaoune, S., Jamal, A., Ronga, D., & Černý, J. (2024). Mitigation of Detrimental Effects of Salinity on Sweet Pepper through Biochar-Based Fertilizers Derived from Date Palm Wastes. *Phyton*, *93*(11), 2993–3011. DOI: 10.32604/phyton.2024.057536

Mushtaq, T., Bano, A., & Ullah, A. (2025). Effects of Rhizospheric Microbes, Growth Regulators, and Biochar in Modulating Antioxidant Machinery of Plants Under Stress. *Journal of Plant Growth Regulation*, *44*(5), 1846–1867. DOI: 10.1007/s00344-024-11548-9

Palansooriya, K. N., Wong, J. T. F., Hashimoto, Y., Huang, L., Rinklebe, J., Chang, S. X., Bolan, N., Wang, H., & Ok, Y. S. (2019). Response of microbial communities to biochar-amended soils: A critical review. *Biochar*, *1*(1), 3–22. DOI: 10.1007/s42773-019-00009-2

Rana, M. S., Chen, H., Deng, S., Imran, M., Abdellah, Y. A. Y., Li, W., Lin, J., Li, J., & Wang, R. (2025). Biological insights into alleviating heavy metal toxicity through the simultaneous supply of biochar biofilters derived from the Mikania micrantha and molybdenum nanoparticles. *Biochar*, *7*(1), 23. Advance online publication. DOI: 10.1007/s42773-024-00414-2

Rathinapriya, P., Maharajan, T., Jothi, R., Prabakaran, M., Lee, I. B., Yi, P. H., & Jeong, S. T. (2024). Unlocking biochar impacts on abiotic stress dynamics: A systematic review of soil quality and crop improvement. *Frontiers in Plant Science*, *15*, 1479925. DOI: 10.3389/fpls.2024.1479925

Razzaq, S., Zhou, B., Zia-Ur-rehman, M., Maqsood, M. A., Hussain, S., Bakhsh, G., Zhang, Z., Yang, Q., & Altaf, A. R. (2022). Cadmium Stabilization and Redox Transformation Mechanism in Maize Using Nanoscale Zerovalent-Iron-Enriched Biochar in Cadmium-Contaminated Soil. *Plants*, *11*(8), 1074. DOI: 10.3390/plants11081074

Ren, J., Ren, X., Deng, Z., Zhang, H., Wang, J., Zhang, C., Lu, F., & Shi, J. (2025). Ecological effects of biochar in heavy metal-contaminated soils from multidimensional perspective: Using meta-analysis. *Bioresource Technology*, *432*, 132695. DOI: 10.1016/j.biortech.2025.132695

Sarim, M., Jan, T., Khattak, S. A., Mihoub, A., Jamal, A., Saeed, M. F., Soltani-Gerdefaramarzi, S., Tariq, S. R., Fernández, M. P., Mancinelli, R., & Radicetti, E. (2022). Assessment of the Ecological and Health Risks of Potentially Toxic Metals in Agricultural Soils from the Drosh-Shishi Valley, Pakistan. *Land (Basel)*, *11*(10), 1663. DOI: 10.3390/land11101663

Sinha, S., Basak, A., & Mondol, M. S. A. (2025). Understanding heavy metal accumulation in crops: Sources, plant responses, tolerance mechanisms, and environmental effects. *Journal of Environmental Science and Health. Part C, Toxicology and Carcinogenesis*, *43*(3), 269–294. DOI: 10.1080/26896583.2025.2521207

Touzout, N., Bouchibane, M., Tahraoui, H., Mihoub, A., Zhang, J., Amrane, A., Ahmad, I., Danish, S., Alahmadi, T. A., & Ansari, M. J. (2024c). Silicon-mediated resilience: Unveiling the protective role against combined cypermethrin and hymexazol phytotoxicity in tomato seedlings. *Journal of Environmental Management, 369*, 122370. DOI: 10.1016/j.jenvman.2024.122370

Touzout, N., Mehallah, H., Moralent, R., Moulay, M., & Nemmiche, S. (2021b). Phytotoxic evaluation of neonicotinoid imidacloprid and cadmium alone and in combination on tomato (Solanum lycopersicum L.). *Ecotoxicology (London, England), 30*(6), 1126–1137. DOI: 10.1007/s10646-021-02421-6

Touzout, N., Mehallah, H., Moralent, R., Nemmiche, S., & Benkhelifa, M. (2021a). Co-contamination of deltamethrin and cadmium induce oxidative stress in tomato plants (Solanum lycopersicum L.). *Acta Physiologiae Plantarum, 43*(6), 1–10. DOI: 10.1007/s11738-021-03261-x

Touzout, N., Mihoub, A., Ahmad, I., Jamal, A., & Danish, S. (2024a). Deciphering the role of nitric oxide in mitigation of systemic fungicide induced growth inhibition and oxidative damage in wheat. *Chemosphere, 364*, 143046. DOI: 10.1016/j.chemosphere.2024.143046

Touzout, N., Mihoub, A., Boukheddimi, M., Moualdia, A. O., Ahmad, I., Jamal, A., Danish, S., Alarfaj, A. A., Alharbi, S. A., & Javed Ansari, M. (2024b). Nitric oxide application alleviates fungicide and ampicillin co-exposure induced phytotoxicity by regulating antioxidant defense, detoxification system, and secondary metabolism in wheat seedlings. *Journal of Environmental Management, 372*, 123337. DOI: 10.1016/j.jenvman.2024.123337

Touzout, N., Mihoub, A., Krache, F., Ali, E., & Ahmad, R. (2025). Biochar Improves Growth Traits, Photosynthetic Pigments, Antioxidant Enzymes, and Metabolites Accumulation of Wheat (Triticum aestivum L) Under Nickel Toxicity. *Journal of Soil Science and Plant Nutrition, 25*(3), 1–16. DOI: 10.1007/s42729-025-02482-0

Ul Haq, I., Ahmed, U., Ali Khan, I., Rehman, A., Saeed, M. F., Alkahtani, J., Elshikh, M. S., Černý, J., & Jamal, A. (2025). Nano zinc-enriched biochar improves spinach growth and soil nutrient dynamics in degraded soils. *Journal of Plant Nutrition*, *48*(15), 2632–2647. DOI: 10.1080/01904167.2025.2487122

Van Nguyen, Q., Tran, H. M. T., Nguyen, H. T., Van Hoang, H., & Nguyen, K. M. (2025). The Effect of pH on Nitrogen and Phosphorus Recovery Efficiency From Wastewater Using Biochar Derived From Corncob. *Water, Air, and Soil Pollution*, *236*(11), 1–13. DOI: 10.1007/s11270-025-08377-1

Weldeslassie, T., Naz, H., Singh, B., & Oves, M. (2017). Chemical contaminants for soil, air and aquatic ecosystem. In *Modern Age Environmental Problems and their Remediation* (pp. 1–22). Springer., DOI: 10.1007/978-3-319-64501-8_1

Wu, R., Wen, Y., Xu, T., Du, Y., Qi, D., Chang, W., Li, K., Ping, Y., Fan, X., & Song, F. (2025). Rhizophagus irregularis combined with biochar enhances Cr tolerance in Ricinus communis by altering host plant's antioxidant and photosynthetic characteristics. *Industrial Crops and Products*, *227*, 120769. DOI: 10.1016/j.indcrop.2025.120769

Wu, Y., Tan, Z., Zhou, T., Wang, Z., Xue, Z., Yan, Y., He, H., & Pan, J. (2025). Regulation mechanisms of biochar-derived endogenous substances on heavy metal mobility in the soil-crop system: A critical review. *Environmental Research*, *285*, 122557. DOI: 10.1016/j.envres.2025.122557

Wu, Y., Wang, X., Zhang, L., Zheng, Y., Liu, X., & Zhang, Y. (2023). The critical role of biochar to mitigate the adverse impacts of drought and salinity stress in plants. *Frontiers in Plant Science*, *14*(May), 1–22. DOI: 10.3389/fpls.2023.1163451

Xu, N., Zhang, N., Yi, P., Chen, L., Dai, H., Zhang, J., Li, W., Li, R., Liu, A., Zhou, Z., & Tu, X. (2024). Integrated physio-biochemistry and RNA-seq revealed the mechanism underlying biochar-mediated alleviation of compound heavy metals (Cd, Pb, As) toxicity in cotton. *Ecotoxicology and Environmental Safety*, *284*, 116974. DOI: 10.1016/j.ecoenv.2024.116974

Yuan, P., Wang, J., Pan, Y., Shen, B., & Wu, C. (2019). Review of biochar for the management of contaminated soil: Preparation, application and prospect. *The Science of the Total Environment*, *659*, 473–490. DOI: 10.1016/j.scitotenv.2018.12.400

Zeeshan, N., Farooqi, Z. U. R., Ahmad, I., Murtaza, G., Jamal, A., Saifullah, , Abdul Qadir, A., & Radicetti, E. (2023). Trace Metals in Rice Grains and Their Associated Health Risks from Conventional and Non-Conventional Rice Growing Areas in Punjab-Pakistan. *Sustainability (Basel)*, *15*(9), 7259. DOI: 10.3390/su15097259

Zhao, S., Zhao, S., & Wang, B. (2025). Combined application of biochar and PGPB on crop growth and heavy metals accumulation: A meta-analysis. *Environmental Pollution*, *381*, 126626. DOI: 10.1016/j.envpol.2025.126626

Zulfiqar, B., Raza, M. A. S., Saleem, M. F., Aslam, M. U., Iqbal, R., Muhammad, F., Amin, J., Ibrahim, M. A., & Khan, I. H. (2022). Biochar enhances wheat crop productivity by mitigating the effects of drought: Insights into physiological and antioxidant defense mechanisms. *PLoS ONE*, *17*(4 April), 1–16. DOI: 10.1371/journal.pone.0267819

Chapter 8

# Nanotechnology Intricates the Tolerance Mechanism of Plants Subjected to Metal Toxicity

**Hafiza Muniba Din Muhammad**
http://orcid.org/0000-0003-1771-3624
*Bahauddin Zakariya University, Multan, Pakistan*

## ABSTRACT

*Abiotic stress conditions occur from climate obstacles to crop productivity. Among these, metal toxicity in progressively increasing and polluting the soils. Current aimed to explore the adverse effects of metals on plants. It explores the sources of these metals, examining both natural occurrences and human-induced activities, and investigates the mechanisms through which plants absorb them. Metal pollution negatively affects plant and microbiome well-being, producing reactive oxygen species (ROS) that harm essential macromolecules which further contributes in various developmental processes. The need for significant development in conventional stress-resistant plant cultivars has prompted the investigation of novel strategies such as phytohormones, biostimulants, organic amendments, and nanotechnology. Nanoparticles have significant impact on alteration of*

DOI: 10.4018/979-8-3693-9899-9.ch008

*morphological and biochemical properties of normal and stressed plants. Long-term ecological studies and customized formulations are key components of the future prospects in nanoparticle strategies for metal toxicity.*

## INTRODUCTION

Metal contamination is becoming a more significant problem as a result of rapid urbanization and industrialization, fertilizer and agrochemical implementation, irrational mining, and waste management (Srivastava et al., 2017). Living things are at serious risk from metal toxicity, mostly as a result of increased human activity that contaminates the environment (Luo et al., 2009). The growth and productivity may be negatively impacted by oxidative stress and other physiological and biochemical changes due to metal stress (Karn et al., 2021). Enhancing resistance in affected organisms requires an understanding of the mechanisms behind heavy metal stress and possible mitigation techniques (Kumar et al., 2017). Metal exposure produces several toxic substances, which cause oxidative damage to plant cells. This could pose a risk to human health by entering the food chain in addition to reducing crop yields and inhibiting plant growth (Rascio and Navari-Izzo, 2011). Metal toxicity has a major detrimental effect on plant growth and metabolism because of higher buildup of ROS and resultant oxidative stress. Crop output and food safety are eventually impacted by the physiological and biochemical disturbances caused by metal toxicity (Cheng, 2003). Metals in access affect cellular functions in plants, which results in stunted growth and decreased biomass (Nagajyoti et al., 2010). Lipid peroxidation and harm to proteins, nucleic acids, and cellular membranes are caused by the buildup of ROS. Enzyme activity can be inhibited by metal toxicity, which can upset metabolic processes vital to plant health (Fryzova et al., 2017). Numerous physiological and biochemical mechanisms that improve plant resilience are involved in the process by which nanoparticles reduce metal toxicity in plants (Viehweger, 2014).

Nanoparticles have promising effects in reducing the negative effects of heavy metals by enhancing plant growth and antioxidant responses (Soni et al., 2024). Nanoparticles can minimize toxicity by decreasing the metals' bioavailability and plant absorption by activating compounds (enzymatic and non-enzymatic) under metal toxicity (Faizan et al., 2023). Nanoparticles increase the activity of both enzymatic and non-enzymatic antioxidants to scavenge reactive oxygen species (ROS) produced under metal toxicity (Chowardhara et al., 2024). Plants have established several homeostatic mechanisms to control the development and uptake of metalloids to mitigate the negative effects of metal toxicity (Zhao et al., 2020). However, the efficacy of metal detoxification needs improvement in stressed plants. Silicon nanoparticles boost the uptake of essential nutrients, including calcium and potassium, which are critical for maintaining cellular integrity and function under metal stress (Rahman et al., 2023). The application of nanoparticles has been linked to enhanced growth metrics, such as increased chlorophyll content and biomass, that are essential for plant health (Weisany et al., 2024). Reducing the amount of bioavailable metal pollutants, controlling the expression of genes related to metal/metalloid transport, restoring the apoplastic barrier's ability to intercept metal contaminants, giving stressed plants more nutrients, strengthening both enzymatic and non-enzymatic antioxidant systems, and boosting the biosynthesis of protective agents like organic acids, compatible solutes, phytochelatin, and root secretions are some important strategies for improving tolerance in metal stressed plants (Ali et al., 2019; Koleva et al., 2022).

One promising way to address a major environmental issue is by using nanoparticles to reduce metal toxicity in plants (Mustafa and Komatsu, 2016). Even in contaminated soils, their special qualities allow plants to interact with metals in several ways by increasing plant resistance and growth (Soni et al., 2024). This response summarizes the results of current research, emphasizing the ways in which nanoparticles reduce the toxicity of metals and enhance plant health. Heavy metals and nanoparticles may combine to produce complexes that decrease the metals' bioavailability and absorption by plants (Jalil and Ansari, 2019). This interaction lessens the negative effects of metal toxicity on plant metabolism by detoxifying the soil (Ghouri et al., 2024). Nanoparticles boost both enzymatic and non-enzymatic antioxidants in plants. This increase in antioxidant capacity promotes development and yield by reducing oxidative injury due to metal toxicity (Masarovičová and Kráľová, 2012). Nanoparticles can lower the mobility of metals by changing physico-chemical characteristics of soil, including pH and microbial activity, strengthening the apoplastic barrier in plants. This alteration prevents metals from entering plant tissues (Faizan et al., 2023).

Nanoparticles have demonstrated synergistic effects in improving phytoremediation when combined with microorganisms that promote plant development (Weisany et al., 2024). For instance, the combination of particular bacterial strains and biogenic iron oxide nanoparticles enhances the growth and chlorophyll content of plants under metal toxicity (Naaz et al., 2023). Because silicon nanoparticles occur naturally and can increase plant resilience, they are especially useful in reducing heavy metal stress. They are essential for detoxifying metals and enhancing plant physiological processes (Yan et al., 2024). Plant-derived biogenic iron oxide nanoparticles have shown great promise in promoting plant development and lowering the toxicity of metals. In stressed plants, the use has been associated with decreased electrolyte leakage and increased antioxidant enzyme activity (Du et al., 2022).

## NANOPARTICLES REGULATE THE INHIBITION OF SEEDLING GROWTH AND PLANT PRODUCTIVITY

A major problem currently affecting agriculture is metal toxicity, which has an impact on plant productivity, photosynthetic efficiency, vegetative growth, and seedling growth. Heavy metals build up in soils that pose a serious threat to crop health and productivity due to industrialization, mining, wastewater irrigation, and overuse of agrochemicals (Asati et al., 2016). Nanoparticles have become a promising tool in recent years to lessen the growth and development-inhibiting effects of metal toxicity (Shaikh et al., 2025). They can interact with plant systems at the cellular and molecular levels due to their small size, high surface area, and improved reactivity, and provide several ways to control plant responses in hazardous environments (Dev et al., 2018). The seedlings' growth is frequently reduced under metal toxicity because of oxidative damage, disruption of cellular metabolism, and suppression of photosynthesis (Singh et al., 2024). ZnO NPs have been shown to lessen cadmium toxicity (). These enzymes shield seedlings from oxidative stress-induced damage to membranes by decreasing Cd absorption and increasing the activity of antioxidant enzymes such as peroxidase (POD), catalase (CAT), and superoxide dismutase (SOD), organelles by neutralizing reactive oxygen species (ROS), which are generally overproduced during metal stress (Akhtar et al., 2021). Furthermore, zinc and iron-based nanoparticles can restore vital micronutrients whose absorption is frequently hindered by metal pollutants, encouraging the establishment and early growth of seedlings (Yousaf et al., 2024). Silicon-based nanoparticles are essential for regulating metal toxicity. Silicon nanoparticles have been shown to lessen the accumulation of metals in plant tissues by encouraging the deposition of silicon in cell walls, which serves as a physical barrier preventing metal entrance and promoting metal immobilization in the rhizosphere (Rajput et al., 2021). In addition to preventing harmful metals from moving to aboveground areas, this shielding effect maintains the integrity of root structures, which are essential for absorbing nutrients and water (Ahmed et al., 2023). Additionally, research indicates that silicon nanoparticles increase the expression of metal transporter genes, which control the internal distribution and compartmentalization of heavy metals and lessen their phytotoxic effects on photosynthesis and metabolic processes (Ashraf et al., 2024).

Nanoparticles regulate several physiological processes. For instance, nanoparticles have been shown to improve the chlorophyll fluorescence, stomatal conductance, and photosynthetic pigments under metal toxicity (Koleva et al., 2022). Maintaining energy production and carbohydrate metabolism, which are closely related to biomass increase and successful reproduction, depends on these advancements (Bhatt et al., 2020). Nanoparticles improve flowering, fruit set, and production by maintaining photosynthetic efficiency even in contaminated soils. The priming of seeds with nanoparticles has demonstrated promising results in enhancing seed vigor and germination when exposed to metal toxicity (Zeb et al., 2022). Before exposure to metal stress, seed nano-priming may trigger antioxidant pathways and precondition the seed metabolism. This pre-activation allows the seedlings to better survive early stress, ensuring a more uniform and stronger establishment in hazardous soils (Zhu et al., 2019). However, care must be taken when using nanoparticles in agriculture. The efficiency and possible phytotoxicity of nanoparticles are greatly influenced by their concentration, size, coating, and type (Yasin et al., 2024). Higher doses may be harmful because of excessive ROS formation or metal-nanoparticle interactions, even though low to moderate concentrations frequently have positive effects (Aqeel et al., 2022). Therefore, for nanoparticles to be used safely, it is crucial to learn the crosstalk among metal toxicity, nanoparticles, and plant response. A promising tactic to combat the suppression of plant productivity and seedling growth caused by metal toxicity is the use of nanoparticles (Nawazet al., 2023). They are useful instruments in sustainable agriculture because of their potential to boost antioxidant defense, restrict metal uptake, increase nutrient availability, and maintain photosynthesis (Souri et al., 2021). Impact of nanoparticles on seedlings' growth and plant production under metal toxicity (Table 1).

*Table 1. Impact of nanoparticles on seedlings' growth and plant production under metal toxicity*

| Nanoparticles | Kind of metals | Crop name | Impact on stressed plants | References |
|---|---|---|---|---|
| ZnONPs | Cadmium | Rice | • Boosts plant antioxidant defense system<br>• Enhanced plant biomass<br>• Improved plant yield | Gowayed (2017) |
| $SiO_2$NPs | Arsenic | Mustard | • Excites ROS scavenging<br>• Lightens root inhibition<br>• Decline oxidative stress<br>• Higher seedling growth and plant yield | Faisal et al. (2025) |
| $Fe_3O_4$NPs | Chromium | Soybean | • Increases enzymatic activities<br>• Restores seedling growth<br>• Better pods and seeds per plant | Praveen et al. (2018) |
| AgNPs | Nickel | Mung bean | • Activates antioxidant response<br>• Higher germination rate and seedling growth<br>• Enhanced shoot size and seed weight | Sharma et al. (2024) |
| CuONPs | Zinc | Barley | • Triggers detox paths<br>• Elevated root growth<br>• Excellent fruit set<br>• Superior yield | Karmous et al. (2023) |

ZnONPs = zinc oxide nanoparticles; $SiO_2$NPs = silicon dioxide nanoparticles; $Fe_3O_4$NPs = iron nanoparticles; AgNPs = silver nanoparticles; CuONPs = copper oxide nanoparticles; ROS = reactive oxygen species

## NANOPARTICLES NORMALIZE THE INHIBITION OF PHOTOSYNTHESIS IN PLANTS

Plant growth and development are severely hampered by metal toxicity, which is brought on by toxic quantities of certain elements (Bertrand and Poirier, 2005). This is mainly because it interferes with the photosynthetic system. These metals hinder photosynthesis by destroying chloroplast structures, preventing chlorophyll biosynthesis, and causing an excess of ROS, which oxidize proteins and lipids (Syed et al., 2018). However, recent developments in nanotechnology have brought attention to the possibility of nanoparticles as a practical means of reducing these phytotoxic effects and enabling stressed plants to resume their regular photosynthetic functions (Mendoza-Soto et al., 2012). Nanoparticles can enter plant tissues and trigger physiological reactions that reduce metal toxicity because of their special physicochemical characteristics, which include small size, high surface area, and reactivity (Yruela, 2013). Nanoparticles have been shown in numerous studies to be important in restoring photosynthetic activity under metal stress. These nanoparticles reduce metal toxicity by strengthening the antioxidant defense system, encouraging photosynthesis, controlling metal translocation, and maintaining chloroplast structures (Tighe-Neira et al., 2018). Nanoparticles counteract metal-induced photosynthetic suppression through various important processes, such as the decrease of oxidative stress (Faizan et al., 2023). Heavy metals cause an overabundance of ROS, including hydroxyl radicals, hydrogen peroxide, and superoxide radicals, which harm chlorophyll molecules and interfere with electron transport in photosystems I and II. Antioxidant enzymes such as CAT, POD, and SOD have been demonstrated to be activated by NPs. This increased antioxidant activity shields the photosynthetic machinery from oxidative damage by scavenging ROS. Under metal toxicity, plant maintain higher chlorophyll concentration and a more stable photosynthetic pigment composition with nanoparticles (Khalid et al., 2022). Furthermore, nanoparticles support the preservation of chloroplast ultrastructure, which is often disturbed by exposure to metal toxicity. Metals can disorganize the thylakoid membranes and chloroplasts to enlarge. The nanoparticles ensure appropriate light absorption and electron transport by restoring grana lamellae and maintaining the integrity of the chloroplast membrane

(Verma et al., 2024). Nanoparticles dramatically increased net photosynthetic rate as well as chlorophyll fluorescence characteristics of rice plants under cadmium stress, underscoring their protective effect in preserving functional chloroplasts (Roy et al., 2025). Moreover, heavy metal absorption and compartmentalization are influenced by nanoparticles. For instance, silicon nanoparticles reduce the mobility and translocation of harmful metal ions to aerial portions by forming complexes with them in the rhizosphere or inside plant tissues (Djanaguiraman et al., 2018). This localized immobilization reduces direct interference with photosynthesis by keeping metals from passing through chloroplasts (Zhang et al., 2025). Zinc oxide nanoparticles help the plant produce chlorophyll and activate photosynthetic enzymes like Rubisco by providing vital micronutrients and reducing the toxicity of lead and cadmium (Sharma et al., 2025). Genes linked to the photosynthetic pathway have been shown to have their expression altered by nanoparticles. Under stress, several nanoparticles cause the overexpression of genes linked to photosynthesis and counteract the downregulation caused by metals (Aamer et al., 2025). These transcriptional alterations promote biomass accumulation, energy generation, and $CO_2$ absorption. Nanoparticles offer a possible means of reducing the effects of metal toxicity by restoring normal photosynthesis in plants (Rani et al., 2025). They resist metal-induced damage in several ways, including gene expression modification, metal sequestration, antioxidant activation, and chloroplast protection (Hassan et al., 2024). However, substance concentration, application technique, and plant species all have a significant impact on the use of nanoparticles (Kang et al., 2024). Hence, nanotechnology presents a promising strategy for improving plant resistance to environmental stresses such as metal toxicity (Hussain et al., 2024). Nanoparticles regulate the photosynthetic system of metal-stressed plants (Figure 1).

*Figure 1. Nanoparticles regulate photosynthetic system of metal stressed plants*

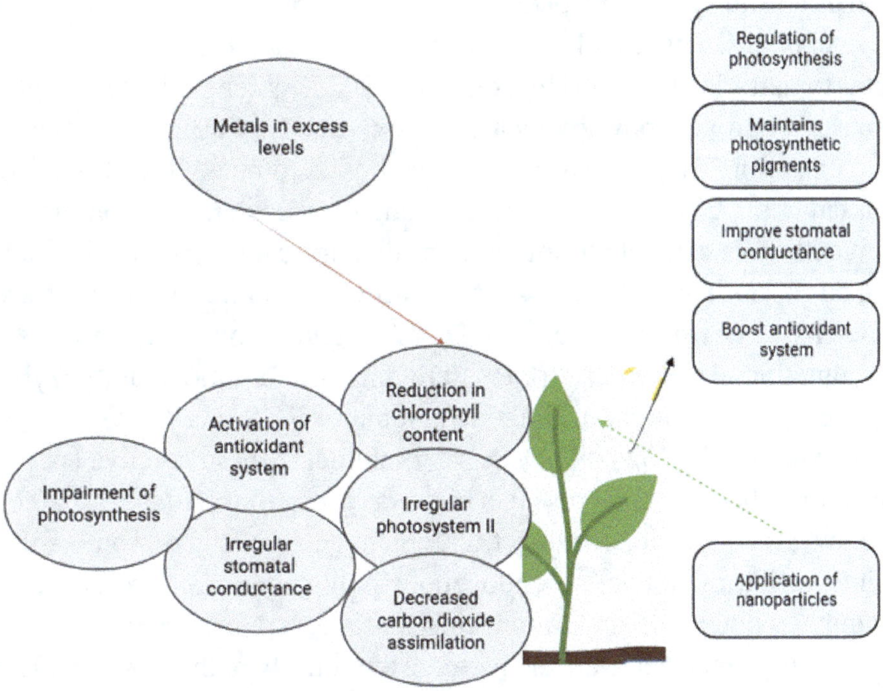


## NANOPARTICLES STRENGTHEN THE ANTIOXIDANT SYSTEM OF PLANTS

Nanotechnology has led to an excellent revolution in the cultivation of crops in polluted soils (Freire et al., 2024). Nanoparticles had the potential to strengthen the antioxidant system of metal-stressed plants. Researchers are developing novel methods that improve plant productivity and tolerance by utilizing the special qualities of nanoparticles (Song et al., 2026). The uses, workings, and possible drawbacks of nanotechnology in stress management were explored by several plant researchers (Sharma et al., 2019). Several studies have been carried out to understand the basic working mechanism of nanoparticles for reducing metal toxicity. The morpho-physiological and biochemical characteristics of plants are affected by metal toxicity

(Kukde et al., 2019). Plant growth-regulating nanoparticles may be harmful to plants when present in high concentrations (Kumari et al., 2024). Numerous strategies have been used to counteract the stress that metals cause in plants. These include lowering the bioavailable concentration of metal toxicity in the soil (Hudson et al., 2025), strengthening the plant defenses, improving physiological processes, developing protective agents, and controlling the expression of genes linked to heavy metal transport (Hassan et al., 2022). Moreover, chelating organic acids that build up in plant cell walls can help to mitigate the damage caused by metal toxicity in plants (Ghorbani et al., 2024). Nanoparticles can improve the synthesis of defensive chemicals. SOD, POD, CAT, glutathione reductase (GR), and glutathione peroxidase (GPX) are among the antioxidant enzymes that are essential for controlling the amount of ROS in cells (Gaur et al., 2021; Altaf et al., 2024). They serve to eradicate these reactive species, especially when they are present in high concentrations (Altaf et al., 2023). Moreover, nanoparticles interact with plant cell walls and promote their disintegration by activating endogenous polyphenols, antioxidants, and vitamins (C and E) (Li and Imran, 2024). Moreover, these chemicals relax the structure of the cell wall and increase the activity of the enzymes. They lessen oxidative stress by scavenging ROS produced during nanoparticle dissemination (Rai et al., 2021). Nanoparticles can enter plant tissues more deeply while preserving cellular homeostasis, attributable to their dual activity (Zhao et al., 2024). Nanoparticles boost the antioxidant defense system of metal-stressed plants (Figure 2).

*Figure 2. Nanoparticles boost antioxidant defense system of metal stressed plants*

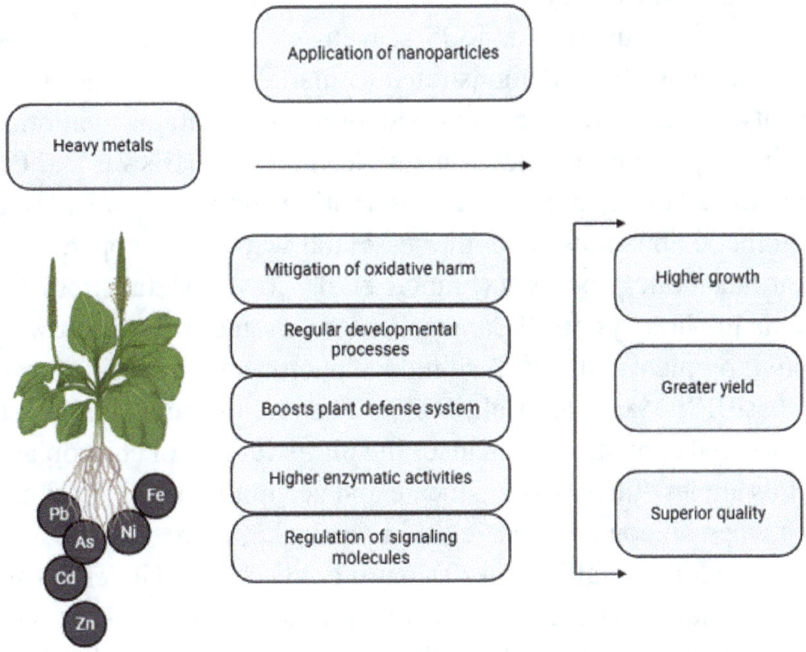


## NANOPARTICLES MAINTAIN METABOLIC ACTIVITIES IN PLANTS

Nanoparticles can interact with plant systems because of their special physicochemical characteristics, which include a high surface area-to-volume ratio, tunable reactivity, and controlled release behavior (Munyaneza et al., 2025). Metal and metal oxide nanoparticles have been shown to play an important role in reducing metabolic dysfunctions caused by metal toxicity (Ahmad et al., 2024; El-Mogy et al., 2025). These nanoparticles reduce metal stress by controlling antioxidant defense systems, stabilizing photosynthetic pigments, improving nutrient balance, and altering signaling pathways essential for plant growth (Javed et al., 2024). Hydrogen peroxide ($H_2O_2$), superoxide anion ($O_2^-$), and hydroxyl radicals are produced in large quantities under metal toxicity, which causes oxidative damage to proteins and cellular membranes (Chen, 2018). By increasing the activity of enzyme antioxidants such as SOD, CAT, and POD, nanoparticles, particularly those of zinc oxide (ZnO), titanium oxide ($TiO_2$), and cerium oxide ($CeO_2$), function as ROS scavengers either directly or indirectly. $TiO_2$-NPs have been demonstrated to markedly increase SOD and CAT activity, which lowers lipid peroxidation and maintains membrane integrity in the presence of cadmium and lead toxicity (Biswas and Pal, 2024; Muhammad et al., 2024). The preservation and improvement of the photosynthetic apparatus is another essential way that nanoparticles sustain plant metabolic processes (Francis et al., 2024). Metal stress frequently results in photosystem II damage, chlorosis, and the breakdown of chlorophyll pigments, all of which reduce photosynthetic efficiency (Rastogi et al., 2017; Manzoor et al., 2025). The use of nanoparticles stabilizes chlorophyll content and enhances the ultrastructure of chloroplasts. Toxic metals are less likely to translocate and accumulate in photosynthetic tissues when silicon nanoparticles, in particular, increase cellular integrity and form a physical barrier (Tripathi et al., 2016). Under stress, this in turn maintains the energy metabolism and carbon assimilation process. Moreover, nanoparticles have the ability to control the distribution and absorption of vital nutrients, which are frequently disrupted by hazardous metals' competitive inhibition (Hussain et al., 2020). Nanoparticles sustain the nitrogen metabolism necessary for protein and enzyme synthesis by improving the absorption of magnesium, iron, and nitrogen in addition to reducing the buildup of lead and cadmium (Cui et al., 2017). Moreover, increased nutritional availability promotes the production of osmolytes, which are essential for preserving the osmotic equilibrium of cells and protecting metabolic enzymes from stress. These osmolytes include proline, carbohydrates, and amino acids. Moreover, signaling molecules that control stress-responsive gene expression and metabolic adaptations are modulated by nanoparticles (Manzoor et al., 2023). For instance, nitric oxide-releasing nanoparticles have demonstrated in enhancing plant resistance to metal toxicity by altering redox homeostasis and initiating antioxidant pathways (Ali et al., 2019). Nevertheless, nanoparticles' advantageous effects are mainly based on their concentration. Excessive nanoparticles can result in phytotoxicity and metabolic disruptions, whereas lower quantities typically activate protective mechanisms and preserve metabolic function

(David et al., 2024). To achieve positive results, it is crucial to optimize the type, size, concentration, and application technique of nanoparticles. Nanoparticles have a great deal of promise for maintaining plant metabolic processes under metal toxicity (Fatemi et al., 2021). Nanoparticles promote plant tolerance by controlling signaling cascades, photosynthetic activity, nutritional balance, and antioxidant systems (Mousavi, 2022; Naz et al., 2022; Muhammad et al., 2025). To comprehend the long-term effects of nanoparticle application, sustainable nano-enabled methods for enhancing plant resistance to environmental pollutants require more research. Nanoparticles' contribution to the regulation of metabolic activities of plants against metal toxicity (Table 2).

*Table 2. Nanoparticles' contribution to the regulation of metabolic activities of plants against metal toxicity*

| Metabolic activities | Toxic effects | Nanoparticles impact | References |
|---|---|---|---|
| Antioxidant enzyme activity | Excess of ROS causes oxidative injury | Boost SOD, CAT, POD, APX, and GR to scavenge toxic ROS | Kukde et al. (2019 |
| Nitrogen metabolism | Restriction of nitrate reductase and inhibition of ammonium assimilation | Higher nitrogen assimilation and nitrate reductase movement | Rawat et al. (2018) |
| Photosynthesis | Distraction of chlorophyll content formation | Improve chlorophyll content, greater photopigments, and better ion transport | Feigl (2023) |
| Water relations and osmotic balance | Interrupted water uptake and turgor pressure in stressed plants | Regulate water uptake, relative water content, and osmoprotectant buildup | Karmous et al. (2023) |
| Lipid metabolism | Higher lipid peroxidation and membrane rupturing | Diminish MDA and $H_2O_2$. Protect membrane integrity | Francis et al. (2024) |
| Mineral nutrient uptake | Inhibited mineral assimilation in stressed plants | Improve minerals uptake | Gaur et al. (2021) |

continued on following page

*Table 2. Continued*

| Metabolic activities | Toxic effects | Nanoparticles impact | References |
|---|---|---|---|
| Carbohydrate metabolism | Poor sugar formation and energy distribution | Greater sugar levels and better starch metabolism | Kumari et al. (2024) |
| Secondary metabolism | Variation in secondary metabolites production | Activation of phenolics, ascorbic acid, and flavonoids to eliminate metals | Souri et al. (2021) |

ROS = reactive oxygen species, SOD = superoxide dismutase, CAT = catalases, POD = peroxidase, APX = ascorbate, and GR = glutathione reductase, MDA = malondialdehyde, hydrogen peroxide ($H_2O_2$).

## CONCLUSION

Global crop production endured various climatic and stress-related difficulties. Heavy metals are not biodegradable and can stay in the soil for extended periods of time, posing a risk to human health and soil microbes. However, heavy metal pollution and remediation techniques have drawn a lot of attention globally. Several techniques have been effectively used to increase plants' ability to grow in soils contaminated with metals by either tolerating or accumulating extra metal in plant tissues. Nanotechnology has emerged as a crucial element for sustainable development in order to mitigate the harmful impacts of metal toxicity. The best and most cost-effective strategy recently developed is the addition of nanoparticles to plants to tolerate metal toxicity and to remediate these harmful elements. Their capacity to permeate plants across a wide surface area, more efficient adsorption, and targeted distribution allow them to assist in controlling photosynthesis and detoxifying toxic substances, which improves crop growth, yield, and seed germination. Nanoparticles also aid in the healthy growth of plants by regulating the mobile and immobile forms of metals. The beneficial effects of applying nanoparticles, particularly their ability to increase metal tolerance and boost plant development, suggest that their use in remediating metal-polluted soils.

## REFERENCES


Aamer, E., Saeed, M., Liaqat, F., Khan, M. R., Naeem, M. K., Mustafa, G., & Malik, R. N. (2025). Unraveling the potential of zinc oxide nanoparticles on alleviating arsenic toxicity in wheat (Triticum aestivum L.) varieties: A comparative assessment of histology and phenomics. *Physiology and Molecular Biology of Plants : an International Journal of Functional Plant Biology*, *31*(10), 1–20. DOI: 10.1007/s12298-025-01590-6 PMID: 41164116

Ahmad, R., Muhammad, H. M. D., Naz, S., Tiwari, R. K., Lal, M. K., Ahmad, P., & Altaf, M. A. (2024). From stress to success: Silicon and nano-silicon strategies for enhancing fruit yield and quality in cadmium-exposed trees. *South African Journal of Botany*, *167*, 441–447. DOI: 10.1016/j.sajb.2024.02.042

Ahmed, S., Iqbal, M., Ahmad, Z., Iqbal, M. A., Artyszak, A., Sabagh, A. E., Alharby, H. F., & Hossain, A. (2023). Foliar application of silicon-based nanoparticles improve the adaptability of maize (Zea mays L.) in cadmium contaminated soils. *Environmental Science and Pollution Research International*, *30*(14), 41002–41013. DOI: 10.1007/s11356-023-25189-0 PMID: 36626058

Akhtar, N., Khan, S., Rehman, S. U., Rehman, Z. U., Khatoon, A., Rha, E. S., & Jamil, M. (2021). Synergistic effects of zinc oxide nanoparticles and bacteria reduce heavy metals toxicity in rice (Oryza sativa L.) plant. *Toxics*, *9*(5), 113. DOI: 10.3390/toxics9050113 PMID: 34065355

Ali, S., Rizwan, M., Hussain, A., Zia ur Rehman, M., Ali, B., Yousaf, B., Wijaya, L., Alyemeni, M. N., & Ahmad, P. (2019). Silicon nanoparticles enhanced the growth and reduced the cadmium accumulation in grains of wheat (Triticum aestivum L.). *Plant Physiology and Biochemistry : PPB*, *140*, 1–8. DOI: 10.1016/j.plaphy.2019.04.041 PMID: 31078051

Altaf, M. A., Naz, S., Kumar, R., Sardar, H., Nawaz, M. A., Kumar, A., Lal, P., Ahmad, R., Hayat, F., Wani, M. A., Tiwari, R. K., & Lal, M. K. (2023). Unraveling the mechanisms of cadmium toxicity in horticultural plants: Implications for plant health. *South African Journal of Botany*, *163*, 433–442. DOI: 10.1016/j.sajb.2023.10.064

Altaf, M. A., Shahid, R., Naz, S., Ahmad, R., Manzoor, M. A., Alsahli, A. A., Altaf, M. M., & Ahmad, P. (2024). Vanadium toxicity was alleviated by supplementation of silicon in tomato seedlings: Upregulating antioxidative enzymes and glyoxalase system. *Plant Physiology and Biochemistry : PPB*, *213*, 108809. DOI: 10.1016/j.plaphy.2024.108809 PMID: 38875778

Aqeel, U., Aftab, T., Khan, M. M. A., Naeem, M., & Khan, M. N. (2022). A comprehensive review of impacts of diverse nanoparticles on growth, development and physiological adjustments in plants under changing environment. *Chemosphere*, *291*, 132672. DOI: 10.1016/j.chemosphere.2021.132672 PMID: 34756946

Asati, A., Pichhode, M., & Nikhil, K. (2016). Effect of heavy metals on plants: An overview. *International Journal of Application or Innovation in Engineering & Management*, *5*(3), 56–66.

Ashraf, H., Ghouri, F., Liang, J., Xia, W., Zheng, Z., Shahid, M. Q., & Fu, X. (2024). Silicon Dioxide Nanoparticles-based amelioration of Cd toxicity by regulating antioxidant activity and photosynthetic parameters in a line developed from wild rice. *Plants*, *13*(12), 1715. DOI: 10.3390/plants13121715 PMID: 38931146

Bertrand, M., & Poirier, I. (2005). Photosynthetic organisms and excess of metals. *Photosynthetica*, *43*(3), 345–353. DOI: 10.1007/s11099-005-0058-2

Bhatt, D., Bhatt, M. D., Nath, M., Dudhat, R., Sharma, M., & Bisht, D. S. 2020. Application of nanoparticles in overcoming different environmental stresses. *Protective chemical agents in the amelioration of plant abiotic stress: Biochemical and molecular perspectives*, pp.635-654.

Biswas, A., & Pal, S. (2024). Plant-nano interactions: A new insight of nano-phytotoxicity. *Plant Physiology and Biochemistry : PPB*, *210*, 108646. DOI: 10.1016/j.plaphy.2024.108646 PMID: 38657549

Chen, H. (2018). Metal based nanoparticles in agricultural system: Behavior, transport, and interaction with plants. *Chemical Speciation and Bioavailability, 30*(1), 123–134. DOI: 10.1080/09542299.2018.1520050

Cheng, S. (2003). Effects of heavy metals on plants and resistance mechanisms. *Environmental Science and Pollution Research International, 10*(4), 256–264. DOI: 10.1065/espr2002.11.141.2 PMID: 12943010

Chowardhara, B., Saha, B., Awasthi, J. P., Deori, B. B., Nath, R., Roy, S., Sarkar, S., Santra, S. C., Hossain, A., & Moulick, D. (2024). An assessment of nanotechnology-based interventions for cleaning up toxic heavy metal/metalloid-contaminated agroecosystems: Potentials and issues. *Chemosphere, 359*, 142178. DOI: 10.1016/j.chemosphere.2024.142178 PMID: 38704049

Cui, J., Liu, T., Li, F., Yi, J., Liu, C., & Yu, H. (2017). Silica nanoparticles alleviate cadmium toxicity in rice cells: Mechanisms and size effects. *Environmental Pollution, 228*, 363–369. DOI: 10.1016/j.envpol.2017.05.014 PMID: 28551566

David, O. A., Labulo, A. H., Hassan, I., Olawuni, I., Oseghale, C. O., Terna, A. D., Ajayi, O. O., Ayegbusi, S. A., & Owolabi, M. O. (2024). Complexation and immobilization of arsenic in maize using green synthesized silicon nanoparticles (SiNPs). *Scientific Reports, 14*(1), 6176. DOI: 10.1038/s41598-024-56924-3 PMID: 38486015

Dev, A., Srivastava, A. K., & Karmakar, S. (2018). Nanomaterial toxicity for plants. *Environmental Chemistry Letters, 16*(1), 85–100. DOI: 10.1007/s10311-017-0667-6

Djanaguiraman, M., Nair, R., Giraldo, J. P., & Prasad, P. V. V. (2018). Cerium oxide nanoparticles decrease drought-induced oxidative damage in sorghum leading to higher photosynthesis and grain yield. *ACS Omega, 3*(10), 14406–14416. DOI: 10.1021/acsomega.8b01894 PMID: 30411067

Du, J., Liu, B., Zhao, T., Xu, X., Lin, H., Ji, Y., Li, Y., Li, Z., Lu, C., Li, P., Zhao, H., Li, Y., Yin, Z., & Ding, X. (2022). Silica nanoparticles protect rice against biotic and abiotic stresses. *Journal of Nanobiotechnology, 20*(1), 197. DOI: 10.1186/s12951-022-01420-x PMID: 35459250

El-Mogy, M.M., Jan, M.F., Naz, I., Ahmad, I., Ahmad, R. & Altaf, M.T., (2025). Nanoparticle Innovations for Mitigating Metal Toxicity in Plants. *Phyton (0031-9457), 94*(3).

Faisal, M., Bilmez Ozcinar, A., Karadeniz, E., Faizan, M., Sultan, H., & Alatar, A. A. (2025). Supplementation of silicon oxide nanoparticles mitigates the damaging effects of arsenic stress on photosynthesis, antioxidant mechanism and nitrogen metabolism in Brassica juncea. *Scientific Reports, 15*(1), 21476. DOI: 10.1038/s41598-025-04553-9 PMID: 40595861

Faizan, M., Alam, P., Rajput, V. D., Tonny, S. H., Yusuf, M., Sehar, S., Adil, M. F., & Hayat, S. (2023). Nanoparticles: An emerging soil crops saviour under drought and heavy metal stresses. *Egyptian Journal of Soil Science, 63*(3), 355–366. DOI: 10.21608/ejss.2023.220619.1616

Fatemi, H., Esmaiel Pour, B., & Rizwan, M. (2021). Foliar application of silicon nanoparticles affected the growth, vitamin C, flavonoid, and antioxidant enzyme activities of coriander (Coriandrum sativum L.) plants grown in lead (Pb)-spiked soil. *Environmental Science and Pollution Research International, 28*(2), 1417–1425. DOI: 10.1007/s11356-020-10549-x PMID: 32839908

Feigl, G. (2023). The impact of copper oxide nanoparticles on plant growth: A comprehensive review. *Journal of Plant Interactions, 18*(1), 2243098. DOI: 10.1080/17429145.2023.2243098

Francis, D. V., Abdalla, A. K., Mahakham, W., Sarmah, A. K., & Ahmed, Z. F. (2024). Interaction of plants and metal nanoparticles: Exploring its molecular mechanisms for sustainable agriculture and crop improvement. *Environment International, 190*, 108859. DOI: 10.1016/j.envint.2024.108859 PMID: 38970982

Freire, B. M., Lange, C. N., Cavalcanti, Y. T., Monteiro, L. R., Pieretti, J. C., Seabra, A. B., & Batista, B. L. (2024). The dual effect of selenium nanoparticles in rice seedlings: From increasing antioxidant activity to inducing oxidative stress. *Plant Stress, 11*, 100372. DOI: 10.1016/j.stress.2024.100372

Fryzova, R., Pohanka, M., Martinkova, P., Cihlarova, H., Brtnicky, M., Hladky, J. and Kynicky, J., 2017. Oxidative stress and heavy metals in plants. *Reviews of Environmental Contamination and Toxicology, 245*, 129-156.

Gaur, V. K., Sharma, P., Gaur, P., Varjani, S., Ngo, H. H., Guo, W., Chaturvedi, P., & Singhania, R. R. (2021). Sustainable mitigation of heavy metals from effluents: Toxicity and fate with recent technological advancements. *Bioengineered*, *12*(1), 7297–7313. DOI: 10.1080/21655979.2021.1978616 PMID: 34569893

Ghorbani, A., Emamverdian, A., Pehlivan, N., Zargar, M., Razavi, S. M., & Chen, M. (2024). Nano-enabled agrochemicals: Mitigating heavy metal toxicity and enhancing crop adaptability for sustainable crop production. *Journal of Nanobiotechnology*, *22*(1), 91. DOI: 10.1186/s12951-024-02371-1 PMID: 38443975

Ghouri, F., Sarwar, S., Sun, L., Riaz, M., Haider, F. U., Ashraf, H., Lai, M., Imran, M., Liu, J., Ali, S., Liu, X., & Shahid, M. Q. (2024). Silicon and iron nanoparticles protect rice against lead (Pb) stress by improving oxidative tolerance and minimizing Pb uptake. *Scientific Reports*, *14*(1), 5986. DOI: 10.1038/s41598-024-55810-2 PMID: 38472251

Gowayed, S. (2017). Impact of zinc oxide nanoparticles on germination and antioxidant system of maize (Zea mays L.) seedling under cadmium stress. *Journal of Plant Protection Sciences*, *6*(1), 1–11. DOI: 10.21608/jpps.2017.7389

Hassan, M. U., Nawaz, M., Mahmood, A., Shah, A. A., Shah, A. N., Muhammad, F., Batool, M., Rasheed, A., Jaremko, M., Abdelsalam, N. R., Hasan, M. E., & Qari, S. H. (2022). The role of zinc to mitigate heavy metals toxicity in crops. *Frontiers in Environmental Science*, *10*, 990223. DOI: 10.3389/fenvs.2022.990223

Hassan, U. M., Huang, G., Haider, F. U., Khan, T. A., Noor, M. A., Luo, F., Zhou, Q., Yang, B., Ul Haq, M. I., & Iqbal, M. M. (2024). Application of zinc oxide nanoparticles to mitigate cadmium toxicity: Mechanisms and future prospects. *Plants*, *13*(12), 1706. DOI: 10.3390/plants13121706 PMID: 38931138

Hudson, L. G., Dashner-Titus, E. J., & MacKenzie, D. (2025). Zinc as a mechanism-based strategy for mitigation of metals toxicity. *Current Environmental Health Reports*, *12*(1), 5. DOI: 10.1007/s40572-025-00474-x PMID: 39827326

Hussain, B., Lin, Q., Hamid, Y., Sanaullah, M., Di, L., Khan, M. B., He, Z., & Yang, X. (2020). Foliage application of selenium and silicon nanoparticles alleviates Cd and Pb toxicity in rice (Oryza sativa L.). *The Science of the Total Environment*, *712*, 136497. DOI: 10.1016/j.scitotenv.2020.136497 PMID: 31945526

Hussain, M., Kaousar, R., Haq, S. I. U., Shan, C., Wang, G., Rafique, N., Shizhou, W., & Lan, Y. (2024). Zinc-oxide nanoparticles ameliorated the phytotoxic hazards of cadmium toxicity in maize plants by regulating primary metabolites and antioxidants activity. *Frontiers in Plant Science*, *15*, 1346427. DOI: 10.3389/fpls.2024.1346427 PMID: 38304740

Jalil, S. U., & Ansari, M. I. 2019. Nanoparticles and abiotic stress tolerance in plants: synthesis, action, and signaling mechanisms. *Plant signaling molecules*, pp.549-561.

Javed, R., Khan, B., Sharafat, U., Bilal, M., Galagedara, L., Abbey, L., & Cheema, M. (2024). Dynamic interplay of metal and metal oxide nanoparticles with plants: Influencing factors, action mechanisms, and assessment of stimulatory and inhibitory effects. *Ecotoxicology and Environmental Safety*, *271*, 115992. DOI: 10.1016/j.ecoenv.2024.115992 PMID: 38262092

Kang, M., Liu, Y., Weng, Y., Wang, H., & Bai, X. (2024). A critical review on the toxicity regulation and ecological risks of zinc oxide nanoparticles to plants. *Environmental Science. Nano*, *11*(1), 14–35. DOI: 10.1039/D3EN00630A

Karmous, I., Gammoudi, N., & Chaoui, A. (2023). Assessing the potential role of zinc oxide nanoparticles for mitigating cadmium toxicity in Capsicum annuum L. under in vitro conditions. *Journal of Plant Growth Regulation*, *42*(2), 719–734. DOI: 10.1007/s00344-022-10579-4

Karn, R., Ojha, N., Abbas, S., & Bhugra, S. (2021, June). A review on heavy metal contamination at mining sites and remedial techniques. [). IOP Publishing.]. *IOP Conference Series. Earth and Environmental Science*, *796*(1), 012013. DOI: 10.1088/1755-1315/796/1/012013

Khalid, M. F., Iqbal Khan, R., Jawaid, M. Z., Shafqat, W., Hussain, S., Ahmed, T., Rizwan, M., Ercisli, S., Pop, O. L., & Alina Marc, R. (2022). Nanoparticles: The plant saviour under abiotic stresses. *Nanomaterials (Basel, Switzerland)*, *12*(21), 3915. DOI: 10.3390/nano12213915 PMID: 36364690

Koleva, L., Umar, A., Yasin, N. A., Shah, A. A., Siddiqui, M. H., Alamri, S., Riaz, L., Raza, A., Javed, T., & Shabbir, Z. (2022). Iron oxide and silicon nanoparticles modulate mineral nutrient homeostasis and metabolism in cadmium-stressed Phaseolus vulgaris. *Frontiers in Plant Science*, *13*, 806781. DOI: 10.3389/fpls.2022.806781 PMID: 35386669

Kukde, S., Sarangi, B. K., & Purohit, H. (2019). Antioxidant role of nanoparticles for enhancing ecological performance of plant system. [). Elsevier.]. *Comprehensive Analytical Chemistry*, *87*, 159–187. DOI: 10.1016/bs.coac.2019.09.004

Kumar, M., Gogoi, A., Kumari, D., Borah, R., Das, P., Mazumder, P., & Tyagi, V. K. (2017). Review of perspective, problems, challenges, and future scenario of metal contamination in the urban environment. *Journal of Hazardous, Toxic and Radioactive Waste*, *21*(4), 04017007. DOI: 10.1061/(ASCE)HZ.2153-5515.0000351

Kumari, A., Gupta, A. K., Sharma, S., Jadon, V. S., Sharma, V., Chun, S. C., & Sivanesan, I. (2024). Nanoparticles as a tool for alleviating plant stress: Mechanisms, implications, and challenges. *Plants*, *13*(11), 1528. DOI: 10.3390/plants13111528 PMID: 38891334

Li, Q., & Imran, . (2024). Mitigation strategies for heavy metal toxicity and its negative effects on soil and plants. *International Journal of Phytoremediation*, *26*(9), 1439–1452. DOI: 10.1080/15226514.2024.2327611 PMID: 38494751

Luo, Y., Wu, L., Liu, L., Han, C., & Li, Z. 2009. Heavy metal contamination and remediation in Asian agricultural land. In *MARCO symposium*.

Manzoor, M., Korolev, K. and Ahmad, R., 2025. Zinc Oxide Nanoparticles: Abiotic Stress Tolerance in Fruit Crops Focusing on Sustainable Production. *Phyton (0031-9457), 94*(5).

Manzoor, N., Ali, L., Al-Huqail, A. A., Alghanem, S. M. S., Al-Haithloul, H. A. S., Abbas, T., Chen, G., Huan, L., Liu, Y., & Wang, G. (2023). Comparative efficacy of silicon and iron oxide nanoparticles towards improving the plant growth and mitigating arsenic toxicity in wheat (Triticum aestivum L.). *Ecotoxicology and Environmental Safety*, *264*, 115382. DOI: 10.1016/j.ecoenv.2023.115382 PMID: 37619453

Masarovičová, E., & Kráľová, K. (2012). *Plant-heavy metal interaction: phytoremediation, biofortification and nanoparticles. Advances in selected plant physiology aspects*. InTech.

Mendoza-Soto, A. B., Sánchez, F., & Hernández, G. (2012). MicroRNAs as regulators in plant metal toxicity response. *Frontiers in Plant Science*, *3*, 105. DOI: 10.3389/fpls.2012.00105 PMID: 22661980

Mousavi, S. M. (2022). Silicon and nano-silicon mediated heavy metal stress tolerance in plants. In *Silicon and nano-silicon in environmental stress management and crop quality improvement* (pp. 181–191). Academic Press. DOI: 10.1016/B978-0-323-91225-9.00012-1

Muhammad, H.M.D., Naz, S., Bibi, Z., Ahmed, M.A., Stanciu, A.S. & Ahmad, R. (2025). Nanoparticles and Phytohormonal Synergy in Plants: Sustainable Agriculture Approach. *Phyton (0031-9457), 94*(9).

Muhammad, H. M. D., Naz, S., Lal, M. K., Tiwari, R. K., Ahmad, R., Nawaz, M. A., Das, R., & Altaf, M. A. (2024). Melatonin in business with abiotic stresses in vegetable crops. *Scientia Horticulturae*, *324*, 112594. DOI: 10.1016/j.scienta.2023.112594

Munyaneza, V., Zhang, W., Haider, S., Ren, L., Ali, A., Kant, S., & Ding, G. (2025). Mitigating Metal Toxicity in Plants Using Nanoparticles: Mechanisms and Implications for Sustainable Agriculture. *Journal of Agricultural and Food Chemistry*, *73*(28), 17374–17386. DOI: 10.1021/acs.jafc.5c04693 PMID: 40601963

Mustafa, G., & Komatsu, S. (2016). Toxicity of heavy metals and metal-containing nanoparticles on plants. *Biochimica et Biophysica Acta (BBA)-. Biochimica et Biophysica Acta. Proteins and Proteomics, 1864*(8), 932–944. DOI: 10.1016/j.bbapap.2016.02.020

Naaz, H., Rawat, K., Saffeullah, P., & Umar, S. (2023). Silica nanoparticles synthesis and applications in agriculture for plant fertilization and protection: A review. *Environmental Chemistry Letters, 21*(1), 539–559. DOI: 10.1007/s10311-022-01515-9

Nagajyoti, P. C., Lee, K. D., & Sreekanth, T. V. M. (2010). Heavy metals, occurrence and toxicity for plants: A review. *Environmental Chemistry Letters, 8*(3), 199–216. DOI: 10.1007/s10311-010-0297-8

Naz, S., Anjum, M. A., Sadiq, B., Ahmad, R., Altaf, M. A., El-Sheikh, M. A., & Shakoor, A. (2022). Purification of sewage wastewater though sand column filter for lessening of heavy metals accumulation in lettuce, carrot, and cauliflower. *Water (Basel), 14*(22), 3770. DOI: 10.3390/w14223770

Praveen, A., Khan, E., Ngiimei, D. S., Perwez, M., Sardar, M., & Gupta, M. (2018). Iron oxide nanoparticles as nano-adsorbents: A possible way to reduce arsenic phytotoxicity in Indian mustard plant (Brassica juncea L.). *Journal of Plant Growth Regulation, 37*(2), 612–624. DOI: 10.1007/s00344-017-9760-0

Rahman, S., Ahmad, I., & Nafees, M. (2023). Mitigation of heavy metal stress in maize (Zea mays L.) through application of silicon nanoparticles. *Biocatalysis and Agricultural Biotechnology, 50*, 102757. DOI: 10.1016/j.bcab.2023.102757

Rai, G. K., Bhat, B. A., Mushtaq, M., Tariq, L., Rai, P. K., Basu, U., Dar, A. A., Islam, S. T., Dar, T. U., & Bhat, J. A. (2021). Insights into decontamination of soils by phytoremediation: A detailed account on heavy metal toxicity and mitigation strategies. *Physiologia Plantarum, 173*(1), 287–304. DOI: 10.1111/ppl.13433 PMID: 33864701

Rajput, V. D., Minkina, T., Feizi, M., Kumari, A., Khan, M., Mandzhieva, S., Sushkova, S., El-Ramady, H., Verma, K. K., Singh, A., Hullebusch, E. D. V., Singh, R. K., Jatav, H. S., & Choudhary, R. (2021). Effects of silicon and silicon-based nanoparticles on rhizosphere microbiome, plant stress and growth. *Biology (Basel)*, *10*(8), 791. DOI: 10.3390/biology10080791 PMID: 34440021

Rani, S., Kumari, N., & Sharma, V. (2025). Zinc Oxide Nanoparticles Improve Photosynthesis by Modulating Antioxidant System and psb A Gene Expression Under Arsenic Stress in Different Cultivars of Vigna radiata. *BioNanoScience*, *15*(2), 242. DOI: 10.1007/s12668-025-01858-x

Rascio, N., & Navari-Izzo, F. (2011). Heavy metal hyperaccumulating plants: How and why do they do it? And what makes them so interesting? *Plant Science : An International Journal of Experimental Plant Biology*, *180*(2), 169–181. DOI: 10.1016/j.plantsci.2010.08.016 PMID: 21421358

Rastogi, A., Zivcak, M., Sytar, O., Kalaji, H. M., He, X., Mbarki, S., & Brestic, M. (2017). Impact of metal and metal oxide nanoparticles on plant: A critical review. *Frontiers in Chemistry*, *5*, 78. DOI: 10.3389/fchem.2017.00078 PMID: 29075626

Rawat, S., Pullagurala, V. L., Hernandez-Molina, M., Sun, Y., Niu, G., Hernandez-Viezcas, J. A., Peralta-Videa, J. R., & Gardea-Torresdey, J. L. (2018). Impacts of copper oxide nanoparticles on bell pepper (Capsicum annum L.) plants: A full life cycle study. *Environmental Science. Nano*, *5*(1), 83–95. DOI: 10.1039/C7EN00697G

Roy, T. K., Islam, M. S., Mahiddin, N. A., Hossain, S. A., Biswas, T., Antu, U. B., Serity, S. A., Miti, J. F., Akter, S., Roy, S., Biswas, A., Mazrin, M., & Ismail, Z. (2025). Application of nanoparticles (NPs) to ameliorate abiotic stress in economically important crop species: A potential review. *Journal of Crop Health*, *77*(1), 19. DOI: 10.1007/s10343-024-01069-6 PMID: 41649711

Shaikh, A., Jamla, M., Sharma, L., Palawat, B., Pathak, S., Shekhawat, G. S., Tripathi, M., & Kumar, V. (2025). Multifaceted Roles of Biogenic Nanoparticles for Conferring Heavy Metal Tolerance in Plants. *Journal of Crop Health*, *77*(3), 82. DOI: 10.1007/s10343-025-01147-3

Sharma, M. M. M., Kapoor, D., Loyal, A., Kumar, R., Sharma, P., & Husen, A. (2024). *Plant Response to Silver Nanoparticles in Terms of Growth, Development, Production, and Protection: An Overview. Plant Response to Silver Nanoparticles: Plant Growth.* Development, Production, and Protection.

Sharma, S., Raja, V., Bhat, A. H., Kumar, N., Alsahli, A. A., & Ahmad, P. (2025). Innovative strategies for alleviating chromium toxicity in tomato plants using melatonin functionalized zinc oxide nanoparticles. *Scientia Horticulturae, 341*, 113930. DOI: 10.1016/j.scienta.2024.113930

Sharma, S., Singh, V. K., Kumar, A., & Mallubhotla, S. 2019. Effect of nanoparticles on oxidative damage and antioxidant defense system in plants. *Molecular plant abiotic stress: Biology and biotechnology*, pp.315-333.

Singh, R., Choudhary, P., Kumar, S., & Daima, H. K. (2024). Mechanistic approaches for crosstalk between nanomaterials and plants: Plant immunomodulation, defense mechanisms, stress resilience, toxicity, and perspectives. *Environmental Science. Nano, 11*(6), 2324–2351. DOI: 10.1039/D4EN00053F

Song, C., Bibi, Z., Ali, S., & Ahmad, R. (2026). Silicon mitigates aluminum toxicity in lettuce by improving photosynthetic pigments and antioxidant defense system. *Biometals : An International Journal on the Role of Metal Ions in Biology, Biochemistry, and Medicine*, •••, 1–14. DOI: 10.1007/s10534-026-00819-5 PMID: 42050262

Soni, S., Jha, A. B., Dubey, R. S., & Sharma, P. (2024). Mitigating cadmium accumulation and toxicity in plants: The promising role of nanoparticles. *The Science of the Total Environment, 912*, 168826. DOI: 10.1016/j.scitotenv.2023.168826 PMID: 38042185

Souri, Z., Khanna, K., Karimi, N., & Ahmad, P. (2021). Silicon and plants: Current knowledge and future prospects. *Journal of Plant Growth Regulation, 40*(3), 906–925. DOI: 10.1007/s00344-020-10172-7

Srivastava, V., Sarkar, A., Singh, S., Singh, P., De Araujo, A. S., & Singh, R. P. (2017). Agroecological responses of heavy metal pollution with special emphasis on soil health and plant performances. *Frontiers in Environmental Science, 5*, 64. DOI: 10.3389/fenvs.2017.00064

Syed, R.S., Dhriti Kapoor, D.K. and Bhat, A.A., 2018. Heavy metal toxicity in plants: a review.

Tighe-Neira, R., Carmora, E., Recio, G., Nunes-Nesi, A., Reyes-Diaz, M., Alberdi, M., Rengel, Z., & Inostroza-Blancheteau, C. (2018). Metallic nanoparticles influence the structure and function of the photosynthetic apparatus in plants. *Plant Physiology and Biochemistry : PPB*, *130*, 408–417. DOI: 10.1016/j.plaphy.2018.07.024 PMID: 30064097

Tripathi, D. K., Singh, S., Singh, V. P., Prasad, S. M., Chauhan, D. K., & Dubey, N. K. (2016). Silicon nanoparticles more efficiently alleviate arsenate toxicity than silicon in maize cultiver and hybrid differing in arsenate tolerance. *Frontiers in Environmental Science*, *4*, 46. DOI: 10.3389/fenvs.2016.00046

Verma, R., Singh, A., Khare, S., & Kumar, P. (2024). Alleviation of environmental stresses in crop plants by nanoparticles: Recent advances and future perspectives. *Journal of Plant Biochemistry and Biotechnology*, *34*(3), 1–24.

Viehweger, K. (2014). How plants cope with heavy metals. *Botanical Studies*, *55*(1), 35. DOI: 10.1186/1999-3110-55-35 PMID: 28510963

Weisany, W., Razmi, J., Eshaghabadi, A. H., & Pashang, D. (2024). Silicon nanoparticles (SiNP): A novel and sustainable strategy for mitigating environmental stresses in plants. *Journal of Soil Science and Plant Nutrition*, *24*(2), 2167–2191. DOI: 10.1007/s42729-024-01790-1

Yan, G., Huang, Q., Zhao, S., Xu, Y., He, Y., Nikolic, M., Nikolic, N., Liang, Y., & Zhu, Z. (2024). Silicon nanoparticles in sustainable agriculture: Synthesis, absorption, and plant stress alleviation. *Frontiers in Plant Science*, *15*, 1393458. DOI: 10.3389/fpls.2024.1393458 PMID: 38606077

Yasin, M. U., Haider, Z., Munir, R., Zulfiqar, U., Rehman, M., Javaid, M. H., Ahmad, I., Nana, C., Saeed, M. S., Ali, B., & Gan, Y. (2024). The synergistic potential of biochar and nanoparticles in phytoremediation and enhancing cadmium tolerance in plants. *Chemosphere*, *354*, 141672. DOI: 10.1016/j.chemosphere.2024.141672 PMID: 38479680

Yousaf, N., Sardar, M. F., Ishfaq, M., Yu, B., Zhong, Y., Zaman, F., Zhang, F., & Zou, C. (2024). Insights in to iron-based nanoparticles (hematite and magnetite) improving the maize growth (Zea mays L.) and iron nutrition with low environmental impacts. *Chemosphere*, *362*, 142781. DOI: 10.1016/j.chemosphere.2024.142781 PMID: 38972262

Yruela, I. (2013). Transition metals in plant photosynthesis. *Metallomics : Integrated Biometal Science*, *5*(9), 1090–1109. DOI: 10.1039/c3mt00086a PMID: 23739807

Zeb, A., Liu, W., & Zhang, Y. (2022). Effects of nanoparticles on alleviating phytotoxicity of soil heavy metals: potential for enhancing phytoremediation. In *Sustainable Agriculture Reviews 53: Nanoparticles: A New Tool to Enhance Stress Tolerance* (pp. 327–344). Springer International Publishing.

Zhang, X., Zhao, Y., Huang, L., Luo, X., Zhang, C., Mao, Z., Yang, H., & Wang, X. (2025). Zinc oxide nanoparticles alleviated Cd toxicity in Hibiscus syriacus L. by reducing Cd translocation and improving plant growth and root cellular ultrastructure. *Journal of Hazardous Materials*, *491*, 137920. DOI: 10.1016/j.jhazmat.2025.137920 PMID: 40086238

Zhao, D., Wang, P., & Zhao, F. J. (2024). Toxic metals and metalloids in food: Current status, health risks, and mitigation strategies. *Current Environmental Health Reports*, *11*(4), 468–483. DOI: 10.1007/s40572-024-00462-7 PMID: 39352604

Zhao, L., Lu, L., Wang, A., Zhang, H., Huang, M., Wu, H., Xing, B., Wang, Z., & Ji, R. (2020). Nano-biotechnology in agriculture: Use of nanomaterials to promote plant growth and stress tolerance. *Journal of Agricultural and Food Chemistry*, *68*(7), 1935–1947. DOI: 10.1021/acs.jafc.9b06615 PMID: 32003987

Zhu, Y., Xu, F., Liu, Q., Chen, M., Liu, X., Wang, Y., Sun, Y., & Zhang, L. (2019). Nanomaterials and plants: Positive effects, toxicity and the remediation of metal and metalloid pollution in soil. *The Science of the Total Environment*, *662*, 414–421. DOI: 10.1016/j.scitotenv.2019.01.234 PMID: 30690375

# Compilation of References

Aamer, E., Saeed, M., Liaqat, F., Khan, M. R., Naeem, M. K., Mustafa, G., & Malik, R. N. (2025). Unraveling the potential of zinc oxide nanoparticles on alleviating arsenic toxicity in wheat (Triticum aestivum L.) varieties: A comparative assessment of histology and phenomics. *Physiology and Molecular Biology of Plants : an International Journal of Functional Plant Biology*, *31*(10), 1–20. DOI: 10.1007/s12298-025-01590-6 PMID: 41164116

Abdollahi, A., Farsad-Akhtar, N., Mohajel Kazemi, E., & Kolahi, M. (2023). Investigation of the combined effects of cadmium chloride, silver nitrate, lead nitrate, methyl jasmonate, and salicylic acid on morphometric and biochemical characteristics of St. John's wort. *Physiology and Molecular Biology of Plants : an International Journal of Functional Plant Biology*, *29*(2), 173–184. DOI: 10.1007/s12298-023-01281-0

Abdu, N., Abdullahi, A. A., & Abdulkadir, A. (2017). Heavy metals and soil microbes. *Environmental Chemistry Letters*, *15*(1), 65–84. DOI: 10.1007/s10311-016-0587-x

Abideen, Z., Koyro, H. W., Huchzermeyer, B., Ansari, R., Zulfiqar, F., & Gul, B. (2020). Ameliorating effects of biochar on photosynthetic efficiency and antioxidant defence of Phragmites karka under drought stress. *Plant Biology*, *22*(2), 259–266. DOI: 10.1111/plb.13054

Abou Fayssal, S., Kumar, P., Popescu, S. M., Sardar, H., Ahmad, R., Gupta, D., Gaur, S. K., Alharby, H. F., & Al-Ghamdi, A. G. (2024). Health risk assessment of heavy metals in saffron (*Crocus sativus* L.) cultivated in domestic wastewater and lake water irrigated soils. *Heliyon, 10*(5), e27138. DOI: 10.1016/j.heliyon.2024.e27138

Abou Seeda, M., Abou El-Nour, E. Z., Abdallah, M., & El-Bassiouny, H. (2022). Impacts of metal, metalloid and their effects in plant physiology: A review. *Middle East J Agric Res, 11*(3), 838–931.

Adnan, M., Yaseen, T. Hafsa, Khan, M. N., Hayat, K., Wahab, S., Wahid, N., & Kaplan, A. (2025). Use of biostimulants to increase heavy metal tolerance in oilseed crops. In *Oilseed crops under abiotic stress: Mitigation strategies and future perspectives* (pp. 175–197). Springer Nature Singapore.

Adnyana, I. M., Bimantara, P. O., & Roni, N. G. K. (2025). Enhancing Soil Fertility Through Azolla Incorporation: Impacts on Nitrogen Cycling and Cation Exchange Capacity. *Organic Farming, 11*(1), 1–12. DOI: 10.56578/of110101

Ahmad, A., Khan, W. U., Shah, A. A., Yasin, N. A., Naz, S., Ali, A., Tahir, A., & Batool, A. I. (2021). Synergistic effects of nitric oxide and silicon on promoting plant growth, oxidative stress tolerance and reduction of arsenic uptake in Brassica juncea. *Chemosphere, 262*, 128384. DOI: 10.1016/j.chemosphere.2020.128384

Ahmad, B., Jaleel, H., Sadiq, Y. A., Khan, M. M., & Shabbir, A. (2018). Response of exogenous salicylic acid on cadmium induced photosynthetic damage, antioxidant metabolism and essential oil production in peppermint. *Plant Growth Regulation, 86*(2), 273–286. DOI: 10.1007/s10725-018-0427-z

Ahmad, I., Ghaffar, A., Zakir, A., Khan, Z. U. H., Saeed, M. F., Rasool, A., Jamal, A., Mihoub, A., Marzeddu, S., & Boni, M. R. (2022). Activated Biochar Is an Effective Technique for Arsenic Removal from Contaminated Drinking Water in Pakistan. *Sustainability (Basel), 14*(21), 14523. DOI: 10.3390/su142114523

Ahmad, I., Mehmood, S., Muneer, M. A., Jamal, A., Farooq, A., Saeed, M. F., Ali, Q., Daraz, U., Touzout, N., Mihoub, A., & Zhou, J. (2025). Bacillus Inoculation Along with Biogas Slurry Mitigates Nickel Stress in Raphanus Sativus by Regulating Ni Accumulation and the Antioxidant Defense System. *Water, Air, and Soil Pollution, 236*(11), 715. DOI: 10.1007/s11270-025-08307-1

Ahmad, P., Ahanger, M. A., Alyemeni, M. N., Wijaya, L., & Alam, P. (2018). Exogenous application of nitric oxide modulates osmolyte metabolism, antioxidants, enzymes of ascorbate-glutathione cycle and promotes growth under cadmium stress in tomato. *Protoplasma, 255*(1), 79–93. DOI: 10.1007/s00709-017-1132-x

Ahmad, P., Alyemeni, M. N., Wijaya, L., Ahanger, M. A., Ashraf, M., Alam, P., Paray, B. A., & Rinklebe, J. (2021). Nitric oxide donor, sodium nitroprusside, mitigates mercury toxicity in different cultivars of soybean. *Journal of Hazardous Materials, 408*, 124852. DOI: 10.1016/j.jhazmat.2020.124852

Ahmad, P., Raja, V., Ashraf, M., Wijaya, L., Bajguz, A., & Alyemeni, M. N. (2021). Jasmonic acid (JA) and gibberellic acid (GA3) mitigated Cd-toxicity in chickpea plants through restricted cd uptake and oxidative stress management. *Scientific Reports, 11*(1), 19768. DOI: 10.1038/s41598-021-98753-8

Ahmad, R., Muhammad, H. M. D., Naz, S., Tiwari, R. K., Lal, M. K., Ahmad, P., & Altaf, M. A. (2024). From stress to success: Silicon and nano-silicon strategies for enhancing fruit yield and quality in cadmium-exposed trees. *South African Journal of Botany, 167*, 441–447. DOI: 10.1016/j.sajb.2024.02.042

Ahmed, M., Ara, N., Aman, F., Farooq, A., Asghar, M., Muhammad, N., Ahmad, Z., & Bibi, R. (2024). Effect of iron and zinc on growth and yield of chili (*Capsicum annuum* L.). *Journal of Agriculture and Education Research, 2*(2), 1–9.

Ahmed, S., Iqbal, M., Ahmad, Z., Iqbal, M. A., Artyszak, A., Sabagh, A. E., Alharby, H. F., & Hossain, A. (2023). Foliar application of silicon-based nanoparticles improve the adaptability of maize (Zea mays L.) in cadmium contaminated soils. *Environmental Science and Pollution Research International*, *30*(14), 41002–41013. DOI: 10.1007/s11356-023-25189-0 PMID: 36626058

Akhtar, N., Khan, S., Rehman, S. U., Rehman, Z. U., Khatoon, A., Rha, E. S., & Jamil, M. (2021). Synergistic effects of zinc oxide nanoparticles and bacteria reduce heavy metals toxicity in rice (Oryza sativa L.) plant. *Toxics*, *9*(5), 113. DOI: 10.3390/toxics9050113 PMID: 34065355

Alamer, K. H., & Fayez, K. A. (2020). Impact of salicylic acid on the growth and physiological activities of parsley plants under lead toxicity. *Physiology and Molecular Biology of Plants : an International Journal of Functional Plant Biology*, *26*(7), 1361–1373. DOI: 10.1007/s12298-020-00830-1

Alamer, K. H., & Galal, T. M. (2022). Safety assessment and sustainability of consuming eggplant (*Solanum melongena* L.) grown in wastewater-contaminated agricultural soils. *Scientific Reports*, *12*(1), 9768. DOI: 10.1038/s41598-022-13992-7 PMID: 35697742

Alamri, S. A., Siddiqui, M. H., Al-Khaishany, M. Y., Nasir Khan, M., Ali, H. M., Alaraidh, I. A., Alsahli, A. A., Al-Rabiah, H., & Mateen, M. (2018). Ascorbic acid improves the tolerance of wheat plants to lead toxicity. *Journal of Plant Interactions*, *13*(1), 409–419. DOI: 10.1080/17429145.2018.1491067

Alasmary, Z., Majrashi, M. A., Aziz, M. A., & Mumtaz, B. (2025). Combined application of biochar and PGPR alleviates Cd stress in wheat by improving antioxidant defense mechanism and crop physiology. *International Journal of Phytoremediation*, *27*(14), 1–16. DOI: 10.1080/15226514.2025.2531260

Al-Charchafchi, F., & Al-Quadan, F. (2006). Effect of chlorogenic acid on germination and seedling growth, and on the enzymes activity extracted from *Artemisia herba alba* Asso. Part I: Germination and seedling growth. *Dirasat. Pure Sciences, 33*(2), 168–175.

Alengebawy, A., Abdelkhalek, S. T., Qureshi, S. R., & Wang, M. Q. (2021). Heavy metals and pesticides toxicity in agricultural soil and plants: Ecological risks and human health implications. *Toxics, 9*(3), 42. DOI: 10.3390/toxics9030042

Alfonso, E., Andlauer, W., Brück, W. M., & Rienth, M. (2025). Whey protein hydrolysates enhance grapevine resilience to abiotic and biotic stresses. *Frontiers in Plant Science, 16*, 1521275. DOI: 10.3389/fpls.2025.1521275

Ali, M. A., Fahad, S., Haider, I., Ahmed, N., Ahmad, S., Hussain, S., & Arshad, M. 2019. Oxidative stress and antioxidant defense in plants exposed to metal/metalloid toxicity. *Reactive oxygen, nitrogen and sulfur species in plants: production, metabolism, signaling and defense mechanisms*, pp.353-370.

Alikhani, O., & Abbaspour, H. (2019). Effects of methyl jasmonate and cadmium on growth traits, cadmium transport and accumulation, and allene-oxide cyclase gene expression in wheat seedlings. *Revista de Agricultura Neotropical, 6*(3), 20–29. DOI: 10.32404/rean.v6i3.3322

Ali, M. S., & Baek, K.-H. (2020). Jasmonic acid signaling pathway in response to abiotic stresses in plants. *International Journal of Molecular Sciences, 21*(2), 621. DOI: 10.3390/ijms21020621

Ali, S., Rizwan, M., Hussain, A., Zia ur Rehman, M., Ali, B., Yousaf, B., Wijaya, L., Alyemeni, M. N., & Ahmad, P. (2019). Silicon nanoparticles enhanced the growth and reduced the cadmium accumulation in grains of wheat (Triticum aestivum L.). *Plant Physiology and Biochemistry : PPB, 140*, 1–8. DOI: 10.1016/j.plaphy.2019.04.041 PMID: 31078051

Alkharabsheh, H. M., Seleiman, M. F., Battaglia, M. L., Shami, A., Jalal, R. S., Alhammad, B. A., Almutairi, K. F., & Al-Saif, A. M. (2021). Biochar and its broad impacts in soil quality and fertility, nutrient leaching and crop productivity: A review. *Agronomy (Basel)*, *11*(5), 993. Advance online publication. DOI: 10.3390/agronomy11050993

Al-Obaidi, J. R., Jamaludin, A. A., Rahman, N. A., & Ahmad-Kamil, E. I. (2024). How plants respond to heavy metal contamination: A narrative review of proteomic studies and phytoremediation applications. *Planta*, *259*(5), 103. DOI: 10.1007/s00425-024-04378-2

Alsafran, M., Usman, K., Ahmed, B., Rizwan, M., Saleem, M. H., & Al Jabri, H. (2022). Understanding the phytoremediation mechanisms of potentially toxic elements: A proteomic overview of recent advances. *Frontiers in Plant Science*, *13*, 881242. DOI: 10.3389/fpls.2022.881242

Altaf, M. A., Hao, Y., He, C., Mumtaz, M. A., Shu, H., Fu, H., & Wang, Z. (2022). Physiological and biochemical responses of pepper (Capsicum annuum L.) seedlings to nickel toxicity. *Frontiers in Plant Science*, *13*, 950392. DOI: 10.3389/fpls.2022.950392

Altaf, M. A., Naz, S., Kumar, R., Sardar, H., Nawaz, M. A., Kumar, A., Lal, P., Ahmad, R., Hayat, F., Wani, M. A., Tiwari, R. K., & Lal, M. K. (2023). Unraveling the mechanisms of cadmium toxicity in horticultural plants: Implications for plant health. *South African Journal of Botany*, *163*, 433–442. DOI: 10.1016/j.sajb.2023.10.064

Altaf, M. A., Shahid, R., Naz, S., Ahmad, R., Manzoor, M. A., Alsahli, A. A., Altaf, M. M., & Ahmad, P. (2024). Vanadium toxicity was alleviated by supplementation of silicon in tomato seedlings: Upregulating antioxidative enzymes and glyoxalase system. *Plant Physiology and Biochemistry : PPB*, *213*, 108809. DOI: 10.1016/j.plaphy.2024.108809 PMID: 38875778

Altaf, M. M., Awan, Z. A., Ashraf, S., Altaf, M. A., Zhu, Z., Alsahli, A. A., & Ahmad, P. (2024b). Melatonin induced reversibility of vanadium toxicity in muskmelon by regulating antioxidant defense and glyoxalase systems. *Journal of Hazardous Materials*, *473*, 134452. DOI: 10.1016/j.jhazmat.2024.134452

Anbuganesan, V., Vishnupradeep, R., Bruno, L. B., Sharmila, K., Freitas, H., & Rajkumar, M. (2024). Combined Application of Biochar and Plant Growth-Promoting Rhizobacteria Improves Heavy Metal and Drought Stress Tolerance in Zea mays. *Plants, 13*(8), 1143. DOI: 10.3390/plants13081143

Angon, P. B., Islam, M. S., Das, A., Anjum, N., Poudel, A., & Suchi, S. A. (2024). Sources, effects and present perspectives of heavy metals contamination: Soil, plants and human food chain. [Note: Article number added where typical; adjust if exact ID known]. *Heliyon, 10*(7), e28725. DOI: 10.1016/j.heliyon.2024.e28357

Angulo-Bejarano, P. I., Puente-Rivera, J., & Cruz-Ortega, R. (2021). Metal and metalloid toxicity in plants: An overview on molecular aspects. *Plants, 10*(4), 635. DOI: 10.3390/plants10040635

Ansari, M. K. A., Anjum, N. A., Ahmad, A., Umar, S., & Iqbal, M. (2012). Heavy metals in soil and plants: An overview of arsenic, cadmium, chromium and mercury. In N. A. Anjum, S. Umar, & A. Ahmad (Eds.), *Oxidative stress in plants: Causes, consequences and tolerance* (pp. 499–518). I.K. International Publishing House.

Antonangelo, J. A., Sun, X., & Eufrade-Junior, H. de J. (2025). Biochar impact on soil health and tree-based crops: A review. *Biochar, 7*(1), 51. DOI: 10.1007/s42773-025-00450-6

Anwar, S., Shafiq, F., Nisa, Z. U., Usman, U., Ashraf, M. Y., & Ali, N. (2021). Effect of cadmium stress on seed germination, plant growth and hydrolyzing enzymes activities in mungbean seedlings. *Journal of Seed Science, 43*, e202143042. DOI: 10.1590/2317-1545v43256006

Aqeel, U., Aftab, T., Khan, M. M. A., Naeem, M., & Khan, M. N. (2022). A comprehensive review of impacts of diverse nanoparticles on growth, development and physiological adjustments in plants under changing environment. *Chemosphere, 291*, 132672. DOI: 10.1016/j.chemosphere.2021.132672 PMID: 34756946

Arif, M. S., Yasmeen, T., Abbas, Z., Ali, S., Rizwan, M., Aljarba, N. H., Alkahtani, S., & Abdel-Daim, M. M. (2021). Role of exogenous and endogenous hydrogen sulfide (H2S) on functional traits of plants under heavy metal stresses: A recent perspective. *Frontiers in Plant Science, 11*, 545453. DOI: 10.3389/fpls.2020.545453

Arikan-Algul, Y., Mutlu-Durak, H., Kutman, U. B., & Yildiz Kutman, B. (2025). Biostimulant extracts obtained from the brown seaweed *Cystoseira barbata* enhance the growth, yield, quality, and nutraceutical value of soil-grown tomato. *Agronomy (Basel), 15*(5), 1138. DOI: 10.3390/agronomy15051138

Arora, A., Sairam, R. K., & Srivastava, G. C. (2002). Oxidative stress and antioxidative system in plants. *Current Science, 82*(10), 1227–1238.

Arteca, R. N., & Arteca, J. M. (2007). Heavy-metal-induced ethylene production in Arabidopsis thaliana. *Journal of Plant Physiology, 164*(11), 1480–1488. DOI: 10.1016/j.jplph.2006.09.006

Asati, A., Pichhode, M., & Nikhil, K. (2016). Effect of heavy metals on plants: An overview. *International Journal of Application or Innovation in Engineering & Management, 5*(3), 56–66.

Asgher, M., Khan, M. I. R., Anjum, N. A., & Khan, N. A. (2015). Minimising toxicity of cadmium in plants—Role of plant growth regulators. *Protoplasma, 252*(2), 399–413. DOI: 10.1007/s00709-014-0710-4

Ashfaque, F., Farooq, S., Chopra, P., & Chhillar, H. (2020). Improving heavy metal tolerance through plant growth regulators and osmoprotectants in plants. In *Improving Abiotic Stress Tolerance in Plants* (pp. 69–98). CRC Press. DOI: 10.1201/9780429027505-5

Ashraf, H., Ghouri, F., Liang, J., Xia, W., Zheng, Z., Shahid, M. Q., & Fu, X. (2024). Silicon Dioxide Nanoparticles-based amelioration of Cd toxicity by regulating antioxidant activity and photosynthetic parameters in a line developed from wild rice. *Plants, 13*(12), 1715. DOI: 10.3390/plants13121715 PMID: 38931146

Ashraf, M., & Foolad, M. R. (2007). Roles of glycine betaine and proline in improving plant abiotic stress resistance. *Environmental and Experimental Botany*, *59*(2), 206–216. DOI: 10.1016/j.envexpbot.2005.12.006

Asif, R., Yasmin, R., Mustafa, M., Ambreen, A., Mazhar, M., Rehman, A., Umbreen, S., & Ahmad, M. 2022. Phytohormones as plant growth regulators and safe protectors against biotic and abiotic stress. In *Plant Hormones-Recent Advances, New Perspectives and Applications*. IntechOpen. DOI: 10.5772/intechopen.102832

Asija, S., Seth, T., Umar, S., & Gupta, R. (2023). Polyamines and their crosstalk with phytohormones in the regulation of plant defense responses. *Journal of Plant Growth Regulation*, *42*(8), 5224–5246. DOI: 10.1007/s00344-022-10837-5

Askari-Khorasgani, O., & Pessarakli, M. (2019). Phytohormone homeostasis and crosstalk effects in response to osmotic stress. In *Handbook of Plant and Crop Stress* (4th ed., pp. 361–384). CRC Press.

Aslam, M. M., Waseem, M., Jakada, B. H., Okal, E. J., Lei, Z., Saqib, H. S. A., Yuan, W., Xu, W., & Zhang, Q. (2022). Mechanisms of abscisic acid-mediated drought stress responses in plants. *International Journal of Molecular Sciences*, *23*(3), 1084. DOI: 10.3390/ijms23031084

Astier, J., Gross, I., & Durner, J. (2018). Nitric oxide production in plants: An update. *Journal of Experimental Botany*, *69*(14), 3401–3411. DOI: 10.1093/jxb/erx420

Attia, H., & Alamer, K. H. (2024). Supplementation of jasmonic acid mitigates the damaging effects of arsenic stress on growth, photosynthesis and nitrogen metabolism in rice. *Rice (New York, N.Y.)*, *17*(1), 31. DOI: 10.1186/s12284-024-00709-9

Avanci, N. C., Luche, D. D., Goldman, G. H., & Goldman, M. H. D. S. (2010). Jasmonates are phytohormones with multiple functions, including plant defense and reproduction. *Genetics and Molecular Research*, *9*(1), 484–505. DOI: 10.4238/vol9-1gmr754

Bagulho, A., Vilas-Boas, F., Pena, A., Peneda, C., Santos, F. C., Jerónimo, A., de Almeida, R. F., & Real, C. (2015). The extracellular matrix modulates $H_2O_2$ degradation and redox signaling in endothelial cells. *Redox Biology*, *6*, 454–460. DOI: 10.1016/j.redox.2015.09.006 PMID: 26409032

Bahrami, M., Heidari, M., & Ghorbani, H. (2016). Variation in antioxidant enzyme activities, growth and some physiological parameters of bitter melon (Momordica charantia) under salinity and chromium stress. *Journal of Environmental Biology*, *37*(4), 529.

Baker, A. J., McGrath, S. P., Reeves, R. D., & Smith, J. A. C. (2020). Metal hyperaccumulator plants: a review of the ecology and physiology of a biological resource for phytoremediation of metal-polluted soils. *Phytoremediation of contaminated soil and water*, pp.85-107.

Bano, A., Singh, K., Singh, S. P., & Sharma, P. (2023). Abscisic Acid: Metabolism, signaling, and crosstalk with other phytohormones under heavy metal stress. *Stresses*, *3*(4), 665–686. DOI: 10.3390/stresses3040046

Basharat, M. (2019). Water management in the Indus Basin in Pakistan: Challenges and opportunities. In *Indus River Basin* (pp. 375–388). Elsevier. DOI: 10.1016/B978-0-12-812782-7.00017-5

Bashir, F., Mahmooduzzafar, , Siddiqi, T. O., & Iqbal, M. (2007). The antioxidative response system in *Glycine max* (L.) Merr. exposed to deltamethrin, a synthetic pyrethroid insecticide. *Environmental Pollution*, *147*(1), 94–100. DOI: 10.1016/j.envpol.2006.08.013 PMID: 17055627

Bashmakov, D. I., Pynenkova, N. A., Sazanova, K. A., & Lukatkin, A. S. (2012). Effect of the synthetic growth regulator Cytodef and heavy metals on oxidative status in cucumber plants. *Russian Journal of Plant Physiology: a Comprehensive Russian Journal on Modern Phytophysiology*, *59*(1), 59–64. DOI: 10.1134/S1021443712010049

Basit, F., Abbas, S., Sheteiwy, M. S., Bhat, J. A., Alsahli, A. A., & Ahmad, P. (2024). Deciphering the alleviation potential of nitric oxide, for low temperature and chromium stress via maintaining photosynthetic capacity, antioxidant defence, and redox homeostasis in rice (Oryza sativa). *Plant Physiology and Biochemistry : PPB*, *214*, 108957. DOI: 10.1016/j.plaphy.2024.108957

Basit, F., Bhat, J. A., Alyemeni, M. N., Shah, T., & Ahmad, P. (2023). Nitric oxide mitigates vanadium toxicity in soybean (Glycine max L.) by modulating reactive oxygen species (ROS) and antioxidant system. *Journal of Hazardous Materials*, *451*, 131085. DOI: 10.1016/j.jhazmat.2023.131085

Basit, F., Bhat, J. A., Guan, Y., Jan, B. L., Tyagi, A., & Ahmad, P. (2022). Nitric oxide and spermine revealed positive defense interplay for the regulation of the chromium toxicity in soybean (Glycine max L.). *Environmental Pollution*, *308*, 119602. DOI: 10.1016/j.envpol.2022.119602

Basit, F., Khalid, M., Bahadur, A., Bhat, J. A., Ahmad, A., & Ahmad, P. (2025). Alleviation of vanadium toxicity through individual and combined supplementation of spermine and spermidine in soybean: Insights into oxidative stress management, detoxification mechanisms, and root exudates. *South African Journal of Botany*, *184*, 468–479. DOI: 10.1016/j.sajb.2025.06.016

Bayar, J., Ali, N., Dong, Y., Ahmad, U., Anjum, M. M., Khan, G. R., Zaib, M., Jalal, A., Ali, R., & Ali, L. (2024). Biochar-based adsorption for heavy metal removal in water: A sustainable and cost-effective approach. *Environmental Geochemistry and Health*, *46*(11), 428. DOI: 10.1007/s10653-024-02214-w

Bayata, A. (2020). Assessment, accumulation, toxicity and importance of heavy metals in agricultural soil and living system. *American Journal of Environmental Protection*, *9*(6), 121–124. DOI: 10.11648/j.ajep.20200906.11

Begara-Morales, J. C., Chaki, M., Valderrama, R., Sánchez-Calvo, B., Mata-Pérez, C., Padilla, M. N., Corpas, F. J., & Barroso, J. B. (2018). Nitric oxide buffering and conditional nitric oxide release in stress response. *Journal of Experimental Botany*, *69*(14), 3425–3438. DOI: 10.1093/jxb/ery072

Bertrand, M., & Poirier, I. (2005). Photosynthetic organisms and excess of metals. *Photosynthetica, 43*(3), 345–353. DOI: 10.1007/s11099-005-0058-2

Bhatt, D., Bhatt, M. D., Nath, M., Dudhat, R., Sharma, M., & Bisht, D. S. 2020. Application of nanoparticles in overcoming different environmental stresses. *Protective chemical agents in the amelioration of plant abiotic stress: Biochemical and molecular perspectives*, pp.635-654.

Bijekar, S., Padariya, H. D., Yadav, V. K., Gacem, A., Hasan, M. A., Awwad, N. S., Yadav, K. K., Islam, S., Park, S., & Jeon, B. H. (2022). The state of the art and emerging trends in wastewater treatment in developing nations. *Water (Basel), 14*(16), 2537. DOI: 10.3390/w14162537

Bint Tariq, T., & Umer, M. (2025). The potential of seaweed-derived polysaccharides as sustainable biostimulants in agriculture. *International Journal of Biological Macromolecules, 298*, 140009. DOI: 10.1016/j.ijbiomac.2025.140009

Biswas, A., & Pal, S. (2024). Plant-nano interactions: A new insight of nano-phytotoxicity. *Plant Physiology and Biochemistry : PPB, 210*, 108646. DOI: 10.1016/j.plaphy.2024.108646 PMID: 38657549

Biswas, A., & Pal, S. (2025). Role of phytohormones in developing heavy metal stress resilience in rice: Physiological–molecular adjustments, cross-talk, and prospects. *Journal of Plant Growth Regulation, 44*(9), 4960–4980. Advance online publication. DOI: 10.1007/s00344-025-11757-w

Bitarishvili, S., Dikarev, A., Kazakova, E., Bondarenko, E., Prazyan, A., Makarenko, E., Babina, D., Podobed, M., & Geras'kin, S. (2023). Growth, antioxidant system, and phytohormonal status of barley cultivars contrasting in cadmium tolerance. *Environmental Science and Pollution Research International, 30*(21), 59749–59764. DOI: 10.1007/s11356-023-26523-2

Blanco-Canqui, H. (2017). Biochar and Soil Physical Properties. *Soil Science Society of America Journal, 81*(4), 687–711. DOI: 10.2136/sssaj2017.01.0017

Blázquez, M. A. (2024). Polyamines: Their role in plant development and stress. *Annual Review of Plant Biology*, *75*(1), 75. DOI: 10.1146/annurev-arplant-070623-110056

Blokhina, O. B., Virolainen, E., & Fagerstedt, K. V. (2003). Antioxidants, oxidative damage and oxygen deprivation stress: A review. *Annals of Botany*, *91*(2), 179–194. DOI: 10.1093/aob/mcf118 PMID: 12509339

Bücker-Neto, L., Paiva, A. L. S., Machado, R. D., Arenhart, R. A., & Margis-Pinheiro, M. (2017). Interactions between plant hormones and heavy metals responses. *Genetics and Molecular Biology*, *40*(1, suppl 1), 373–386. DOI: 10.1590/1678-4685-gmb-2016-0087

Buet, A., Galatro, A., Ramos-Artuso, F., & Simontacchi, M. (2019). Nitric oxide and plant mineral nutrition: Current knowledge. *Journal of Experimental Botany*, *70*(17), 4461–4476. DOI: 10.1093/jxb/erz129

Buet, A., Moriconi, J. I., Santa-María, G. E., & Simontacchi, M. (2014). An exogenous source of nitric oxide modulates zinc nutritional status in wheat plants. *Plant Physiology and Biochemistry : PPB*, *83*, 337–345. DOI: 10.1016/j.plaphy.2014.08.020

Bulak, P., Walkiewicz, A., & Brzezińska, M. (2014). Plant growth regulators-assisted phytoextraction. *Biologia Plantarum*, *58*(1), 1–8. DOI: 10.1007/s10535-013-0382-5

Buss, W., Belt, K., Hein, Z., Putri, A., Zhu, S., Suwandari, H., & Bentley, A. R. (2025). Hurdles to overcome to achieve biostimulant-driven, low chemical input crop production. *Plants, People, Planet*.

Bystrická, J., Árvay, J., Musilová, J., Vollmannová, A., Tóth, T., & Lenková, M. (2016). The investigation of sensitivity of different types of onion to heavy metal intake from contaminated soil. *International Journal of Environmental Research*, *10*(3), 427–440.

Cabrita, M. T., Duarte, B., Gameiro, C., Godinho, R. M., & Caçador, I. (2018). Photochemical features and trace element substituted chlorophylls as early detection biomarkers of metal exposure in the model diatom *Phaeodactylum tricornutum*. *Ecological Indicators*, *95*, 1038–1052. DOI: 10.1016/j.ecolind.2017.07.057

Carillo, P., Avice, J. C., Vasconcelos, M. W., Du Jardin, P., & Brown, P. H. (2025). Biostimulants in agriculture. [Adjust article ID if needed]. *Physiologia Plantarum*, *177*(1), e70000. DOI: 10.1111/ppl.70046

Carrasco-Castilla, J., Hernández-Álvarez, A. J., Jiménez-Martínez, C., Jacinto-Hernández, C., Alaiz, M., Girón-Calle, J., Vioque, J., & Dávila-Ortiz, G. (2012). Antioxidant and metal chelating activities of peptide fractions from phaseolin and bean protein hydrolysates. *Food Chemistry*, *135*(3), 1789–1795. DOI: 10.1016/j.foodchem.2012.06.016

Castello, D. F., Nejamkin, A., Cassia, R., Correa-Aragunde, N., Fernández, B., Foresi, N., Lombardo, C., Ramirez, L., & Lamattina, L. (2019). The era of nitric oxide in plant biology: Twenty years tying up loose ends. *Nitric Oxide : Biology and Chemistry*, *85*, 17–27. DOI: 10.1016/j.niox.2019.01.013

Cerana, R., & Malerba, M. (2015). Role of nitric oxide in heavy metal stress. In *Nitric oxide action in abiotic stress responses in plants* (pp. 181–192). Springer International Publishing. DOI: 10.1007/978-3-319-17804-2_12

Chai, T. T., Tong, S. R., Law, Y. C., Ismail, N. I. N., Manan, F. A., & Wong, F. C. (2015). Anti-oxidative, metal chelating and radical scavenging effects of protein hydrolysates from blue-spotted stingray. *Tropical Journal of Pharmaceutical Research*, *14*(8), 1349–1355. DOI: 10.4314/tjpr.v14i8.5

Checker, V. G., Kushwaha, H. R., Kumari, P., & Yadav, S. (2018). Role of phytohormones in plant defense: signaling and cross talk. In *Molecular aspects of plant-pathogen interaction* (pp. 159–184). Springer Singapore. DOI: 10.1007/978-981-10-7371-7_7

Chelikani, P., Fita, I., & Loewen, P. C. (2004). Diversity of structures and properties among catalases. *Cellular and Molecular Life Sciences : CMLS*, *61*(2), 192–208. DOI: 10.1007/s00018-003-3206-5 PMID: 14745498

Cheng, C., Liu, J., Liu, J., Gao, Z., Yang, Y., Zhu, B., Yao, F., & Ye, Q. (2025). Selenium alleviates cadmium toxicity in pepper (Capsicum annuum L.) by reducing accumulation, enhancing stress resistance, and promoting growth. *Plants, 14*(9), 1291. DOI: 10.3390/plants14091291

Cheng, S. (2003). Effects of heavy metals on plants and resistance mechanisms. *Environmental Science and Pollution Research International, 10*(4), 256–264. DOI: 10.1065/espr2002.11.141.2 PMID: 12943010

Chen, H. (2018). Metal based nanoparticles in agricultural system: Behavior, transport, and interaction with plants. *Chemical Speciation and Bioavailability, 30*(1), 123–134. DOI: 10.1080/09542299.2018.1520050

Chen, L., Hu, W. F., Long, C., & Wang, D. (2021). Exogenous plant growth regulator alleviate the adverse effects of U and Cd stress in sunflower (Helianthus annuus L.) and improve the efficacy of U and Cd remediation. *Chemosphere, 262*, 127809. DOI: 10.1016/j.chemosphere.2020.127809

Chen, R., Zheng, L., Zhao, J., Ma, J., & Li, X. (2023). Biochar Application Maintains Photosynthesis of Cabbage by Regulating Stomatal Parameters in Salt-Stressed Soil. *Sustainability (Basel), 15*(5), 4206. Advance online publication. DOI: 10.3390/su15054206

Chen, X., Jiang, W., Tong, T., Chen, G., Zeng, F., Jang, S., Gao, W., Li, Z., Mak, M., Deng, F., & Chen, Z. H. (2021). Molecular interaction and evolution of jasmonate signaling with transport and detoxification of heavy metals and metalloids in plants. *Frontiers in Plant Science, 12*, 665842. DOI: 10.3389/fpls.2021.665842

Cheong, J. J., & Choi, D. Y. (2003). Methyl jasmonate as a vital substance in plants. *Trends in Genetics : TIG, 19*(7), 409–413. DOI: 10.1016/S0168-9525(03)00138-0

Choudhary, L. K., & Kumar, A. (2025). Concentration-dependent dual effects of exogenous barium on growth, oxidative stress, and antioxidative enzymes in *Triticum aestivum. Russian Journal of Plant Physiology: a Comprehensive Russian Journal on Modern Phytophysiology, 72*(4), 123. DOI: 10.1134/S1021443724607547

Chowardhara, B., Saha, B., Awasthi, J. P., Deori, B. B., Nath, R., Roy, S., Sarkar, S., Santra, S. C., Hossain, A., & Moulick, D. (2024). An assessment of nanotechnology-based interventions for cleaning up toxic heavy metal/metalloid-contaminated agroecosystems: Potentials and issues. *Chemosphere*, *359*, 142178. DOI: 10.1016/j.chemosphere.2024.142178 PMID: 38704049

Clemens, S. (2006). Toxic metal accumulation, responses to exposure and mechanisms of tolerance in plants. *Biochimie*, *88*(11), 1707–1719. DOI: 10.1016/j.biochi.2006.07.003

Čmiková, N., Galovičová, L., Miškeje, M., Borotová, P., Kluz, M., & Kačániová, M. (2022). Determination of antioxidant, antimicrobial activity, heavy metals and elements content of seaweed extracts. *Plants*, *11*(11), 1493. DOI: 10.3390/plants11111493

Creissen, G. P., Edwards, E. A., & Mullineaux, P. M. (1994). Glutathione reductase and ascorbate peroxidase. In Foyer, C. H., & Mullineaux, P. M. (Eds.), *Causes Of Photoxidative Stress and Amelioration of Defense Systems in Plants* (pp. 343–364). CRC Press.

Cui, J., Liu, T., Li, F., Yi, J., Liu, C., & Yu, H. (2017). Silica nanoparticles alleviate cadmium toxicity in rice cells: Mechanisms and size effects. *Environmental Pollution*, *228*, 363–369. DOI: 10.1016/j.envpol.2017.05.014 PMID: 28551566

DalCorso, G., Manara, A., & Furini, A. (2013). An overview of heavy metal challenge in plants: From roots to shoots. *Metallomics : Integrated Biometal Science*, *5*(9), 1117–1132. DOI: 10.1039/c3mt00038a

Dalvi, A. A., & Bhalerao, S. A. (2013). Response of plants towards heavy metal toxicity: An overview of avoidance, tolerance and uptake mechanism. *Annals of Plant Sciences*, *2*(9), 362–368.

Dar, N. A., Amin, I., Wani, W., Wani, S. A., Shikari, A. B., Wani, S. H., & Masoodi, K. Z. (2017). Abscisic acid: A key regulator of abiotic stress tolerance in plants. *Plant Gene*, *11*, 106–111. DOI: 10.1016/j.plgene.2017.07.003

Dar, S. A., Lone, R., Tyub, S., Kamili, A. N., & Nawchoo, I. A. (2023). Crosstalk of ethylene and salicylic acid in the amelioration of toxic effects of heavy metal stress in mustard. In *Plant Phenolics in Abiotic Stress Management* (pp. 173–193). Springer Nature Singapore. DOI: 10.1007/978-981-19-6426-8_9

Dash, L., Singh, A., & Roychoudhury, A. (2025). Non-enzymatic antioxidant defense systems in plants. In *Role of Antioxidants in Mitigating Plant Stress* (pp. 51–61). Academic Press. DOI: 10.1016/B978-0-443-26799-4.00013-X

David, O. A., Labulo, A. H., Hassan, I., Olawuni, I., Oseghale, C. O., Terna, A. D., Ajayi, O. O., Ayegbusi, S. A., & Owolabi, M. O. (2024). Complexation and immobilization of arsenic in maize using green synthesized silicon nanoparticles (SiNPs). *Scientific Reports*, *14*(1), 6176. DOI: 10.1038/s41598-024-56924-3 PMID: 38486015

Demidchik, V. (2015). Mechanisms of oxidative stress in plants: From classical chemistry to cell biology. *Environmental and Experimental Botany*, *109*, 212–228. DOI: 10.1016/j.envexpbot.2014.06.021

Derrick, A., Yohana, M. A., Yudong, Z., Li, G., Tan, B., & Zhang, S. (2025). Understanding the detrimental effects of heavy metal pollution in shrimp farming and treatment methods—A review. *Annals of Animal Science*, *25*(1), 35–56. DOI: 10.2478/aoas-2024-0041

Dev, A., Srivastava, A. K., & Karmakar, S. (2018). Nanomaterial toxicity for plants. *Environmental Chemistry Letters*, *16*(1), 85–100. DOI: 10.1007/s10311-017-0667-6

Dhalaria, R., Kumar, D., Kumar, H., Nepovimova, E., Kuča, K., Torequl Islam, M., & Verma, R. (2020). Arbuscular mycorrhizal fungi as potential agents in ameliorating heavy metal stress in plants. *Agronomy (Basel)*, *10*(6), 815. DOI: 10.3390/agronomy10060815

Dhiman, S., Khanna, K., Kumar, P., Bhardwaj, T., Devi, K., Sharma, N., Sharma, P., Arora, P., Kapoor, N., Sharma, A., & Bhardwaj, R. (2025). Divulging molecular perspectives of plant defense machinery under heavy metal toxicity. *Journal of Plant Growth Regulation*, *44*(4), 1352–1388. DOI: 10.1007/s00344-023-11138-1

Diaz-Chuquizuta, H., Coral-Cardenas, S. A., Arévalo-Aranda, Y. G., Sánchez-Ojanasta, M., Diaz-Chuquizuta, P., Ocaña-Reyes, J. A., Solórzano Acosta, R. A., & Cuevas Giménez, J. P. (2025). Inoculation of Bacillus subtilis in acidic soil amended with biochar and liming materials in maize cultivation. *Scientific Reports*, *15*(1), 28678. DOI: 10.1038/s41598-025-13230-w

Dickin, S. K., Schuster-Wallace, C. J., Qadir, M., & Pizzacalla, K. (2016). A review of health risks and pathways for exposure to wastewater use in agriculture. *Environmental Health Perspectives*, *124*(7), 900–909. DOI: 10.1289/ehp.1509995 PMID: 26824464

Djanaguiraman, M., Nair, R., Giraldo, J. P., & Prasad, P. V. V. (2018). Cerium oxide nanoparticles decrease drought-induced oxidative damage in sorghum leading to higher photosynthesis and grain yield. *ACS Omega*, *3*(10), 14406–14416. DOI: 10.1021/acsomega.8b01894 PMID: 30411067

Dlamini, S. W., Bertling, I., & Ngcobo, B. L. (2024). Enhancing growth, yield and physiological quality of sweet pepper fruit through application of fertiliser and moringa leaf powder (MLP). *Acta Agriculturæ Scandinavica. Section B, Soil and Plant Science*, *74*(1), 2396973. DOI: 10.1080/09064710.2024.2396973

Duan, X., Li, X., Ding, F., Zhao, J., Guo, A., Zhang, L., Yao, J., & Yang, Y. (2015). Interaction of nitric oxide and reactive oxygen species and associated regulation of root growth in wheat seedlings under zinc stress. *Ecotoxicology and Environmental Safety*, *113*, 95–102. DOI: 10.1016/j.ecoenv.2014.11.030

Dubey, R. S. (2010). Metal toxicity, oxidative stress and antioxidative defense system in plants. In *Reactive oxygen species and antioxidants in higher plants* (pp. 177–203). CRC Press.

Dubey, R. S., Srivastava, R. K., & Pessarakli, M. (2021). Physiological mechanisms of nitrogen absorption and assimilation in plants under stressful conditions. In Pessarakli, M. (Ed.), *Handbook of Plant and Crop Physiology* (pp. 579–616). CRC Press. DOI: 10.1201/9781003093640-36

Dubey, S., Shri, M., Gupta, A., Rani, V., & Chakrabarty, D. (2018). Toxicity and detoxification of heavy metals during plant growth and metabolism. *Environmental Chemistry Letters, 16*(4), 1169–1192. DOI: 10.1007/s10311-018-0741-8

Du, J., Liu, B., Zhao, T., Xu, X., Lin, H., Ji, Y., Li, Y., Li, Z., Lu, C., Li, P., Zhao, H., Li, Y., Yin, Z., & Ding, X. (2022). Silica nanoparticles protect rice against biotic and abiotic stresses. *Journal of Nanobiotechnology, 20*(1), 197. DOI: 10.1186/s12951-022-01420-x PMID: 35459250

Einhardt, A. M., Ferreira, S., & Rodrigues, F. Á. (2021). Biochemical and physiological responses of soybean [*Glycine max* (L.) Merrill] to nickel toxicity. *Bragantia, 80*, Article e1721.

Ekinci, M., Shams, M., Turan, M., Ucar, S., Yaprak, E., Yuksel, E. A., Aydin, M., Ilhan, E., Agar, G., Ercisli, S., & Yildirim, E. (2024). Chitosan mitigated the adverse effect of Cd by regulating antioxidant activities, hormones, and organic acids contents in pepper (*Capsicum annuum* L.). *Heliyon, 10*(17), Article e366xx. (Use exact article ID if available)

El Khattabi, O., El Hasnaoui, S., Toura, M., Henkrar, F., Collin, B., Levard, C., Colin, F., Merghoub, N., Smouni, A., & Fahr, M. (2023). Seaweed extracts as promising biostimulants for enhancing lead tolerance and accumulation in tomato (*Solanum lycopersicum*). *Journal of Applied Phycology, 35*(1), 459–469. DOI: 10.1007/s10811-022-02849-1

El-Mogy, M.M., Jan, M.F., Naz, I., Ahmad, I., Ahmad, R. & Altaf, M.T., (2025). Nanoparticle Innovations for Mitigating Metal Toxicity in Plants. *Phyton (0031-9457), 94*(3).

El-Mogy, M.M., Jan, M.F., Naz, I., Ahmad, I., Ahmad, R. and Altaf, M.T., (2025). Nanoparticle Innovations for Mitigating Metal Toxicity in Plants. Phyton (0031-9457), 94(3).

El-Mogy, M.M., Jan, M.F., Naz, I., Ahmad, I., Ahmad, R. and Altaf, M.T., 2025. Nanoparticle Innovations for Mitigating Metal Toxicity in Plants. *Phyton (0031-9457), 94*(3).

El-Mogy, M. M., Jan, M. F., Naz, I., Ahmad, I., Ahmad, R., & Altaf, M. T. (2025). Nanoparticle innovations for mitigating metal toxicity in plants. *Phyton, 94*(3), 623–640. DOI: 10.32604/phyton.2025.063763

El-Saadony, M. T., Desoky, E. S. M., El-Tarabily, K. A., AbuQamar, S. F., & Saad, A. M. (2024). Exploiting the role of plant growth-promoting rhizobacteria in reducing heavy metal toxicity of pepper (*Capsicum annuum* L.). *Environmental Science and Pollution Research International, 31*(18), 27465–27484. DOI: 10.1007/s11356-024-32874-1 PMID: 38512572

Emamverdian, A., & Ding, Y. (2017). Heavy metals induced changes on growth, antioxidant enzyme activity, gas exchange parameters and protein structures in *Sasa kongosanensis* f. *Aureo-striatus. Polish Journal of Environmental Studies, 26*(2), 585–592. DOI: 10.15244/pjoes/65367

Emamverdian, A., Ding, Y., Barker, J., Mokhberdoran, F., Ramakrishnan, M., Liu, G., & Li, Y. (2021). Nitric oxide ameliorates plant metal toxicity by increasing antioxidant capacity and reducing Pb and Cd translocation. *Antioxidants, 10*(12), 1981. DOI: 10.3390/antiox10121981

Emamverdian, A., Ding, Y., Mokhberdoran, F., & Xie, Y. (2015). Heavy metal stress and some mechanisms of plant defense response. *TheScientificWorldJournal, 2015*(1), 756120. DOI: 10.1155/2015/756120

Emamverdian, A., Ding, Y., & Xie, Y. (2020). The role of new members of phytohormones in plant amelioration under abiotic stress with an emphasis on heavy metals. *Polish Journal of Environmental Studies, 29*(2), 1009–1020. DOI: 10.15244/pjoes/108687

Engwa, G. A., Ferdinand, P. U., Nwalo, F. N., & Unachukwu, M. N. (2019). Heavy metal toxicity in humans. In *Poisoning in the modern world: New tricks for an old dog* (pp. 77–100). IntechOpen. DOI: 10.5772/intechopen.82511

Ercal, N., Gurer-Orhan, H., & Aykin-Burns, N. (2001). Toxic metals and oxidative stress part I: Mechanisms involved in metal-induced oxidative damage. *Current Topics in Medicinal Chemistry*, *1*(6), 529–539. DOI: 10.2174/1568026013394831

Faisal, M., Bilmez Ozcinar, A., Karadeniz, E., Faizan, M., Sultan, H., & Alatar, A. A. (2025). Supplementation of silicon oxide nanoparticles mitigates the damaging effects of arsenic stress on photosynthesis, antioxidant mechanism and nitrogen metabolism in Brassica juncea. *Scientific Reports*, *15*(1), 21476. DOI: 10.1038/s41598-025-04553-9 PMID: 40595861

Faizan, M., Alam, P., Rajput, V. D., Tonny, S. H., Yusuf, M., Sehar, S., Adil, M. F., & Hayat, S. (2023). Nanoparticles: An emerging soil crops saviour under drought and heavy metal stresses. *Egyptian Journal of Soil Science*, *63*(3), 355–366. DOI: 10.21608/ejss.2023.220619.1616

Farid, M., Shakoor, M. B., Ehsan, S., Ali, S., Zubair, M., & Hanif, M. S. (2013). Morphological, physiological and biochemical responses of different plant species to Cd stress. *International Journal of Chemical and Biochemical Science*, *3*, 53–60.

Farman, S., Mushtaq, A., & Azeem, M. W. (2019). Plant growth regulators (PGRs) and their applications: A review. *International Journal of Chemical and Biochemical Science*, *15*, 94–103.

Farooqi, Z. U. R., Ahmad, I., Abdul Qadir, A., Murtaza, G., Rafiq, S., Jamal, A., Zeeshan, N., Murtaza, B., Javed, W., Radicetti, E., & Mancinelli, R. (2022). Zeolite-Assisted Immobilization and Health Risks of Potentially Toxic Elements in Wastewater-Irrigated Soil under Brinjal (Solanum melongena) Cultivation. *Agronomy (Basel)*, *12*(10), 2433. DOI: 10.3390/agronomy12102433

Farooq, M. A., Gill, R. A., Islam, F., Ali, B., Liu, H., Xu, J., He, S., & Zhou, W. (2016). Methyl jasmonate regulates antioxidant defense and suppresses arsenic uptake in *Brassica napus* L. *Frontiers in Plant Science*, *7*, 468. DOI: 10.3389/fpls.2016.00468

Fatemi, H., Esmaiel Pour, B., & Rizwan, M. (2021). Foliar application of silicon nanoparticles affected the growth, vitamin C, flavonoid, and antioxidant enzyme activities of coriander (Coriandrum sativum L.) plants grown in lead (Pb)-spiked soil. *Environmental Science and Pollution Research International*, *28*(2), 1417–1425. DOI: 10.1007/s11356-020-10549-x PMID: 32839908

Feigl, G. (2023). The impact of copper oxide nanoparticles on plant growth: A comprehensive review. *Journal of Plant Interactions*, *18*(1), 2243098. DOI: 10.1080/17429145.2023.2243098

Francis, D. V., Abdalla, A. K., Mahakham, W., Sarmah, A. K., & Ahmed, Z. F. (2024). Interaction of plants and metal nanoparticles: Exploring its molecular mechanisms for sustainable agriculture and crop improvement. *Environment International*, *190*, 108859. DOI: 10.1016/j.envint.2024.108859 PMID: 38970982

Freire, B. M., Lange, C. N., Cavalcanti, Y. T., Monteiro, L. R., Pieretti, J. C., Seabra, A. B., & Batista, B. L. (2024). The dual effect of selenium nanoparticles in rice seedlings: From increasing antioxidant activity to inducing oxidative stress. *Plant Stress*, *11*, 100372. DOI: 10.1016/j.stress.2024.100372

Freschi, L. (2013). Nitric oxide and phytohormone interactions: Current status and perspectives. *Frontiers in Plant Science*, *4*, 398. DOI: 10.3389/fpls.2013.00398

Fryzova, R., Pohanka, M., Martinkova, P., Cihlarova, H., Brtnicky, M., Hladky, J. and Kynicky, J., (2017). Oxidative stress and heavy metals in plants. *Reviews of environmental contamination and toxicology volume 245*, pp.129-156.

Fryzova, R., Pohanka, M., Martinkova, P., Cihlarova, H., Brtnicky, M., Hladky, J. and Kynicky, J., 2017. Oxidative stress and heavy metals in plants. *Reviews of environmental contamination and toxicology volume 245*, pp.129-156.

Fryzova, R., Pohanka, M., Martinkova, P., Cihlarova, H., Brtnicky, M., Hladky, J. and Kynicky, J., 2017. Oxidative stress and heavy metals in plants. *Reviews of Environmental Contamination and Toxicology, 245*, 129-156.

Gao, Y., Ma, X., Zhang, Z., Wang, X., & Wang, Y. (2024). Brassinolides signaling pathway: Tandem response to plant hormones and regulation under various abiotic stresses. *Horticulture Advances*, *2*(1), 27. DOI: 10.1007/s44281-024-00050-7

Garg, A., Chauhan, P., Kaur, C., Arora, P. K., Garg, S. K., Singh, V. P., Singh, K. P., & Srivastava, A. (2025). Comprehensive heavy metal remediation mechanisms with insights into CRISPR-Cas9 and biochar innovations. *Biodegradation*, *36*(4), 69. DOI: 10.1007/s10532-025-10165-x

Gaur, V. K., Sharma, P., Gaur, P., Varjani, S., Ngo, H. H., Guo, W., Chaturvedi, P., & Singhania, R. R. (2021). Sustainable mitigation of heavy metals from effluents: Toxicity and fate with recent technological advancements. *Bioengineered*, *12*(1), 7297–7313. DOI: 10.1080/21655979.2021.1978616 PMID: 34569893

Ghazouani, H., Ibrahimi, K., Amami, R., Helaoui, S., Boughattas, I., Kanzari, S., Milham, P., Ansar, S., & Sher, F. (2023). Integrative effect of activated biochar to reduce water stress impact and enhance antioxidant capacity in crops. *The Science of the Total Environment*, *905*(September), 166950. DOI: 10.1016/j.scitotenv.2023.166950

Ghorbani, A., Emamverdian, A., Pehlivan, N., Zargar, M., Razavi, S. M., & Chen, M. (2024). Nano-enabled agrochemicals: Mitigating heavy metal toxicity and enhancing crop adaptability for sustainable crop production. *Journal of Nanobiotechnology*, *22*(1), 91. DOI: 10.1186/s12951-024-02371-1 PMID: 38443975

Ghorbani, A., Pishkar, L., Roodbari, N., Pehlivan, N., & Wu, C. (2021). Nitric oxide could allay arsenic phytotoxicity in tomato (Solanum lycopersicum L.) by modulating photosynthetic pigments, phytochelatin metabolism, molecular redox status and arsenic sequestration. *Plant Physiology and Biochemistry : PPB*, *167*, 337–348. DOI: 10.1016/j.plaphy.2021.08.019

Ghouri, F., Sarwar, S., Sun, L., Riaz, M., Haider, F. U., Ashraf, H., Lai, M., Imran, M., Liu, J., Ali, S., Liu, X., & Shahid, M. Q. (2024). Silicon and iron nanoparticles protect rice against lead (Pb) stress by improving oxidative tolerance and minimizing Pb uptake. *Scientific Reports*, *14*(1), 5986. DOI: 10.1038/s41598-024-55810-2 PMID: 38472251

Ghuge, S. A., Nikalje, G. C., Kadam, U. S., Suprasanna, P., & Hong, J. C. (2023). Comprehensive mechanisms of heavy metal toxicity in plants, detoxification, and remediation. *Journal of Hazardous Materials*, *450*, 131039. DOI: 10.1016/j.jhazmat.2023.131039

Giannakoula, A., Therios, I., & Chatzissavvidis, C. (2021). Effect of lead and copper on photosynthetic apparatus in citrus (Citrus aurantium l.) plants. the role of antioxidants in oxidative damage as a response to heavy metal stress. *Plants*, *10*(1), 1–14. DOI: 10.3390/plants10010155

Gil'vanova, I. R., Enikeev, A. R., Stepanov, S. Y., & Rakhmankulova, Z. F. (2012). Involvement of salicylic acid and nitric oxide in protective reactions of wheat under the influence of heavy metals. *Applied Biochemistry and Microbiology*, *48*(1), 90–94. DOI: 10.1134/S000368381201005X

Gill, S. S., Hasanuzzaman, M., Nahar, K., Macovei, A., & Tuteja, N. (2013). Importance of nitric oxide in cadmium stress tolerance in crop plants. *Plant Physiology and Biochemistry : PPB*, *63*, 254–261. DOI: 10.1016/j.plaphy.2012.12.001

Godwin, P. M., Pan, Y., Xiao, H., & Afzal, M. T. (2019). Progress in Preparation and Application of Modified Biochar for Improving Heavy Metal Ion Removal From Wastewater. *Journal of Bioresources and Bioproducts*, *4*(1), 31–42. DOI: 10.21967/jbb.v4i1.180

Goncharuk, E. A., & Zagoskina, N. V. (2023). Heavy metals, their phytotoxicity, and the role of phenolic antioxidants in plant stress responses with focus on cadmium. *Molecules (Basel, Switzerland)*, *28*(9), 3921. DOI: 10.3390/molecules28093921

Gondor, O. K., Pál, M., Darkó, É., Janda, T., & Szalai, G. (2016). Salicylic acid and sodium salicylate alleviate cadmium toxicity to different extents in maize (Zea mays L.). *PLoS One*, *11*(8), e0160157. DOI: 10.1371/journal.pone.0160157

Gong, Y., Zhao, D., & Wang, Q. (2018). An overview of field-scale studies on remediation of soil contaminated with heavy metals and metalloids: Technical progress over the last decade. *Water Research*, *147*, 440–460. DOI: 10.1016/j.watres.2018.10.024

Gowayed, S. (2017). Impact of zinc oxide nanoparticles on germination and antioxidant system of maize (Zea mays L.) seedling under cadmium stress. *Journal of Plant Protection Sciences*, *6*(1), 1–11. DOI: 10.21608/jpps.2017.7389

Grattan, S. R., Díaz, F. J., Pedrero, F., & Vivaldi, G. A. (2015). Assessing the suitability of saline wastewaters for irrigation of *Citrus* spp.: Emphasis on boron and specific-ion interactions. *Agricultural Water Management*, *157*, 48–58. DOI: 10.1016/j.agwat.2015.01.002

Graziano, M., & Lamattina, L. (2005). Nitric oxide and iron in plants: An emerging and converging story. *Trends in Plant Science*, *10*(1), 4–8. DOI: 10.1016/j.tplants.2004.12.004

Greičiūtė, K. and Vasarevičius, S., 2007. Decontamination of heavy-metal polluted soil BY electrokinetic remediation.

Guarino, C., Paura, B., & Sciarrillo, R. (2018). Enhancing phytoextraction of heavy metals at real scale by combining salicaceae trees with microbial consortia. *Frontiers in Environmental Science*, *6*, 137. DOI: 10.3389/fenvs.2018.00137

Gul, A. S., Naveed, F., Ali, M., Ahmad, R., & Saqib, M. (2021). Effect of different wastewater irrigation regimes on growth of mulberry (Morus macroura Miq.). *Erwerbs-Obstbau*, *63*(3), 331–337. DOI: 10.1007/s10341-021-00577-3

Gul-Lalay, , Ullah, S., Shah, S., Jamal, A., Saeed, M. F., Mihoub, A., Zia, A., Ahmed, I., Seleiman, M. F., Mancinelli, R., & Radicetti, E. (2024). Combined Effect of Biochar and Plant Growth-Promoting Rhizbacteria on Physiological Responses of Canola (Brassica napus L.) Subjected to Drought Stress. *Journal of Plant Growth Regulation*, *43*(6), 1814–1832. DOI: 10.1007/s00344-023-11219-1

Guo, B., Liu, C., Liang, Y., Li, N., & Fu, Q. (2019). Salicylic acid signals plant defence against cadmium toxicity. *International Journal of Molecular Sciences*, *20*(12), 2960. DOI: 10.3390/ijms20122960

Gupta, K. J., Kaladhar, V. C., Fitzpatrick, T. B., Fernie, A. R., Møller, I. M., & Loake, G. J. (2022). Nitric oxide regulation of plant metabolism. *Molecular Plant*, *15*(2), 228–242. DOI: 10.1016/j.molp.2021.12.012

Gupta, S., & Seth, C. S. (2021). Salicylic acid alleviates chromium (VI) toxicity by restricting its uptake, improving photosynthesis and augmenting antioxidant defense in Solanum lycopersicum L. *Physiology and Molecular Biology of Plants : an International Journal of Functional Plant Biology*, *27*(11), 2651–2664. DOI: 10.1007/s12298-021-01088-x

Habibi, G. (2012). Exogenous salicylic acid alleviates oxidative damage of barley plants under drought stress. *Acta Biologica Szegediensis*, *56*(1), 57–63.

Hafeez, A., Rasheed, R., Ashraf, M. A., Qureshi, F. F., Hussain, I., & Iqbal, M. (2022). Effect of heavy metals on growth, physiological and biochemical responses of plants. In *Plants and their Interaction to Environmental Pollution: Damage Detection, Adaptation, Tolerance, Physiological and Molecular Responses* (pp. 139–159). Elsevier. DOI: 10.1016/B978-0-323-99978-6.00006-6

Haghighi, M., Kafi, M., Fang, P., & Gui-Xiao, L. (2010). Humic acid decreased hazardous of cadmium toxicity on lettuce (*Lactuca sativa* L.). *Vegetable Crops Research Bulletin*, *72*, 49.

Haiying, T., Shubin, W., Ying, L., Ying, S., & Yasser, S. M. (2022). Biochar: A promising soil amendment to mitigate heavy metals toxicity in plants. *Notulae Botanicae Horti Agrobotanici Cluj-Napoca*, *50*(3), 12778. DOI: 10.15835/nbha50312778

Haouari, C. C., Nasraoui, A. H., Bouthour, D., Houda, M. D., Daieb, C. B., Mnai, J., & Gouia, H. (2012). Response of tomato (Solanum lycopersicon) to cadmium toxicity: Growth, element uptake, chlorophyll content and photosynthesis rate. *African Journal of Plant Science*, *6*(1), 1–7.

Hareem, M., Danish, S., Pervez, M., Irshad, U., Fahad, S., Dawar, K., Alharbi, S. A., Ansari, M. J., & Datta, R. (2024). Optimizing chili production in drought stress: Combining Zn-quantum dot biochar and proline for improved growth and yield. *Scientific Reports*, *14*(1), 1–15. DOI: 10.1038/s41598-024-57204-w

Hasan, M., & Naushin, F. (2026). Plant responses to heavy metals: Mechanisms and adaptations. In *Optimizing plant health under abiotic-stress environments* (pp. 1–28). CRC Press.

Hasan, M., Naushin, F., & Khan, F. A. (2025). Soil symbionts in alleviation of heavy metals-laden sewage sludge-impacts in legumes: Implications for phytoremediation. *Discover Plants*, *2*(1), 103. DOI: 10.1007/s44372-025-00192-6

Hasan, N., Choudhary, S., Naaz, N., & Sharma, N. (2021). The mechanism of heavy metal elements in various biological process and its deteriorate effects on the productivity of different crop plants. *International Journal of Waste Resources*, *11*, 403.

Hashem, H. A., Hassanein, A. A., & Esmail, N. Y. (2018). Nitric oxide enhances the adaptive responses of lupine plants against heavy-metal stress. *Australian Journal of Crop Science*, *12*(12), 1962–1975. DOI: 10.21475/ajcs.18.12.12.p1379

Hassan, A. S. (2019). Inorganic-based pesticides: A review article. *Egyptian Scientific Journal of Pesticides*, *5*, 39–52.

Hassan, M. U., Nawaz, M., Mahmood, A., Shah, A. A., Shah, A. N., Muhammad, F., Batool, M., Rasheed, A., Jaremko, M., Abdelsalam, N. R., Hasan, M. E., & Qari, S. H. (2022). The role of zinc to mitigate heavy metals toxicity in crops. *Frontiers in Environmental Science*, *10*, 990223. DOI: 10.3389/fenvs.2022.990223

Hassan, U. M., Huang, G., Haider, F. U., Khan, T. A., Noor, M. A., Luo, F., Zhou, Q., Yang, B., Ul Haq, M. I., & Iqbal, M. M. (2024). Application of zinc oxide nanoparticles to mitigate cadmium toxicity: Mechanisms and future prospects. *Plants*, *13*(12), 1706. DOI: 10.3390/plants13121706 PMID: 38931138

Hassena, A. B., Zouari, M., Trabelsi, L., Khabou, W., & Zouari, N. (2018). Physiological improvements of young olive tree (*Olea europaea* L. cv. Chetoui) under short-term irrigation with treated wastewater. *Agricultural Water Management*, *207*, 53–58. DOI: 10.1016/j.agwat.2018.05.024

He, T., Chen, L., Wu, Y., Wang, J., Wu, Q., Sun, J., Ding, C., Zhou, T., Chen, L., Jin, A., Li, Y., & Zhu, Q. (2024). Combined Metabolome and Transcriptome Analyses of Maize Leaves Reveal Global Effect of Biochar on Mechanisms Involved in Anti-Herbivory to Spodoptera frugiperda. *Metabolites*, *14*(9), 498. Advance online publication. DOI: 10.3390/metabo14090498

Hewedy, O. A., Elsheery, N. I., Karkour, A. M., Elhamouly, N., Arafa, R. A., Mahmoud, G. A. E., Dawood, M. F. A., Hussein, W. E., Mansour, A., Amin, D. H., Allakhverdiev, S. I., Zivcak, M., & Brestic, M. (2023). Jasmonic acid regulates plant development and orchestrates stress response during tough times. *Environmental and Experimental Botany*, *208*, 105260. DOI: 10.1016/j.envexpbot.2023.105260

He, Y., Fukushige, H., Hildebrand, D. F., & Gan, S. (2002). Evidence supporting a role of jasmonic acid in Arabidopsis leaf senescence. *Plant Physiology*, *128*(3), 876–884. DOI: 10.1104/pp.010843

Hoque, M. N., Tahjib-Ul-Arif, M., Hannan, A., Sultana, N., Akhter, S., Hasanuzzaman, M., Akter, F., Hossain, M. S., Sayed, M. A., Hasan, M. T., Skalicky, M., Li, X., & Brestič, M. (2021). Melatonin modulates plant tolerance to heavy metal stress: Morphological responses to molecular mechanisms. *International Journal of Molecular Sciences*, *22*(21), 11445. DOI: 10.3390/ijms222111445

Hossain, M. A., Piyatida, P., da Silva, J. A. T., & Fujita, M. (2012). Molecular mechanism of heavy metal toxicity and tolerance in plants: Central role of glutathione in detoxification of reactive oxygen species and methylglyoxal and in heavy metal chelation. *Journal of Botany (Egypt)*, *2012*, 872875. DOI: 10.1155/2012/872875

Hotamisligil, G. S., & Davis, R. J. (2016). Cell signaling and stress responses. *Cold Spring Harbor Perspectives in Biology*, *8*(10), a006072. DOI: 10.1101/cshperspect.a006072 PMID: 27698029

Hou, D., O'Connor, D., Igalavithana, A. D., Alessi, D. S., Luo, J., Tsang, D. C., Sparks, D. L., Yamauchi, Y., Rinklebe, J., & Ok, Y. S. (2020). Metal contamination and bioremediation of agricultural soils for food safety and sustainability. *Nature Reviews. Earth & Environment*, *1*(7), 366–381. DOI: 10.1038/s43017-020-0061-y

Howladar, S. M. (2014). A novel *Moringa oleifera* leaf extract can mitigate the stress effects of salinity and cadmium in bean (*Phaseolus vulgaris* L.) plants. *Ecotoxicology and Environmental Safety*, *100*, 69–75. DOI: 10.1016/j.ecoenv.2013.11.022

Hsu, Y. T., & Kao, C. H. (2003). Role of abscisic acid in cadmium tolerance of rice (Oryza sativa L.) seedlings. *Plant, Cell & Environment*, *26*(6), 867–874. DOI: 10.1046/j.1365-3040.2003.01018.x

Huang, H., Liu, B., Liu, L., & Song, S. (2017). Jasmonate action in plant growth and development. *Journal of Experimental Botany*, *68*(6), 1349–1359. DOI: 10.1093/jxb/erw495

Huang, L., Lee, J. Y., Park, Y. K., & Lee, J. (2025). Heavy metals in seaweed: Implications for health benefits, risks, and safety regulations. *Journal of Agriculture and Food Research, 21*, 101830. DOI: 10.1016/j.jafr.2025.101830

Huang, R., Cui, X., Luo, X., Mao, P., Zhuang, P., Li, Y., Li, Y., & Li, Z. (2021). Effects of plant growth regulator and chelating agent on the phytoextraction of heavy metals by Pfaffia glomerata and on the soil microbial community. *Environmental Pollution, 283*, 117159. DOI: 10.1016/j.envpol.2021.117159

Huang, Y., Zhang, G., Wu, F., Chen, J., & Xiao, Y. (2006). Interaction of salinity and cadmium stresses on antioxidant enzymes, sodium, and cadmium accumulation in four barley genotypes. *Journal of Plant Nutrition, 29*(12), 2215–2225. DOI: 10.1080/01904160600974155

Hu, B., Deng, F., Chen, G., Chen, X., Gao, W., Long, L., Xia, J., & Chen, Z. H. (2020). Evolution of abscisic acid signaling for stress responses to toxic metals and metalloids. *Frontiers in Plant Science, 11*, 909. DOI: 10.3389/fpls.2020.00909

Hudson, L. G., Dashner-Titus, E. J., & MacKenzie, D. (2025). Zinc as a mechanism-based strategy for mitigation of metals toxicity. *Current Environmental Health Reports, 12*(1), 5. DOI: 10.1007/s40572-025-00474-x PMID: 39827326

Huihui, Z., Xin, L., Zisong, X., Yue, W., Zhiyuan, T., Meijun, A., Yuehui, Z., Wenxu, Z., Nan, X., & Guangyu, S. (2020). Toxic effects of heavy metals Pb and Cd on mulberry (Morus alba L.) seedling leaves: Photosynthetic function and reactive oxygen species (ROS) metabolism responses. *Ecotoxicology and Environmental Safety, 195*, 110469. DOI: 10.1016/j.ecoenv.2020.110469

Hussain, B., Ashraf, M. N., Abbas, A., Li, J., & Farooq, M. (2021). Cadmium stress in paddy fields: Effects of soil conditions and remediation strategies. *The Science of the Total Environment, 754*, 142188. DOI: 10.1016/j.scitotenv.2020.142188 PMID: 33254942

Hussain, B., Lin, Q., Hamid, Y., Sanaullah, M., Di, L., Khan, M. B., He, Z., & Yang, X. (2020). Foliage application of selenium and silicon nanoparticles alleviates Cd and Pb toxicity in rice (Oryza sativa L.). *The Science of the Total Environment*, *712*, 136497. DOI: 10.1016/j.scitotenv.2020.136497 PMID: 31945526

Hussain, M., Kaousar, R., Haq, S. I. U., Shan, C., Wang, G., Rafique, N., Shizhou, W., & Lan, Y. (2024). Zinc-oxide nanoparticles ameliorated the phytotoxic hazards of cadmium toxicity in maize plants by regulating primary metabolites and antioxidants activity. *Frontiers in Plant Science*, *15*, 1346427. DOI: 10.3389/fpls.2024.1346427 PMID: 38304740

Hu, Z., Fu, Q., Zheng, J., Zhang, A., & Wang, H. (2020). Transcriptomic and metabolomic analyses reveal that melatonin promotes melon root development under copper stress by inhibiting jasmonic acid biosynthesis. *Horticulture Research*, *7*(1), 59. DOI: 10.1038/s41438-020-0293-5

Ilieva, I., & Sainova, I. (2023). Mechanisms of Action of Heavy Metals, Related with Abnormal Protein and Enzyme Activity in Male Infertility Aspect. *Acta Morphologica et Anthropologica*, *30*(1-2), 1–2. DOI: 10.7546/AMA.30.1-2.2023.09

Imran, M., Hussain, S., Iqbal, A., Saleem, M. H., Rehman, N., Mo, Z., Chen, X., & Tang, X. (2023). Nitric oxide confers cadmium tolerance in fragrant rice by modulating physio-biochemical processes, yield attributes, and grain quality traits. *Ecotoxicology and Environmental Safety*, *261*, 115078. DOI: 10.1016/j.ecoenv.2023.115078

Imran, S., Sumi, M. J., Harine, I. J., Paul, N. C., Mahamud, M. A., Rabbi, R. H. M., Brestic, M., & Rhaman, M. S. (2025). Biochar Amendments in Soil: A Sustainable Approach for Mitigating Heavy Metal Stress in Plants. *Phyton*, *94*(4), 1073–1109. DOI: 10.32604/phyton.2025.064046

Jacob, D. E., Jacob, I. D., Daniel, K. S., Gideon, I. K., & Akpan, U. U. (2025). Innovations in Soil Remediation and Land Restoration. In *Environmental Science and Engineering: Vol. Part F513* (pp. 259–313). Springer. DOI: 10.1007/978-3-031-85327-2_9

Jaiswal, A., Verma, A., & Jaiswal, P. (2018). Detrimental effects of heavy metals in soil, plants, and aquatic ecosystems and in humans. [Pages standardized where possible]. *Journal of Environmental Pathology, Toxicology and Oncology : Official Organ of the International Society for Environmental Toxicology and Cancer, 37*(3), 183–197. DOI: 10.1615/JEnvironPatholToxicolOncol.2018025348

Jalil, S. U., & Ansari, M. I. 2019. Nanoparticles and abiotic stress tolerance in plants: synthesis, action, and signaling mechanisms. *Plant signaling molecules*, pp.549-561.

Jang, G., Yoon, Y., & Choi, Y. D. (2020). Crosstalk with jasmonic acid integrates multiple responses in plant development. *International Journal of Molecular Sciences, 21*(1), 305. DOI: 10.3390/ijms21010305

Jan, S., Bhardwaj, R., Sharma, N. R., & Singh, R. (2024). Unraveling the role of plant growth regulators and plant growth promoting rhizobacteria in phytoremediation. *Journal of Plant Growth Regulation, 43*(8), 2471–2487. DOI: 10.1007/s00344-024-11284-0

Javaid, S., uz Zaman, Q., Sultan, K., Riaz, U., Aslam, A., Sharif, S., ... & Ibraheem, S. (2020). 5. Heavy metals stress, mechanism and remediation techniques in rice (Oryza sativa L.): a review. Pure and Applied Biology (PAB), 9(1), 403-426.

Javed, R., Khan, B., Sharafat, U., Bilal, M., Galagedara, L., Abbey, L., & Cheema, M. (2024). Dynamic interplay of metal and metal oxide nanoparticles with plants: Influencing factors, action mechanisms, and assessment of stimulatory and inhibitory effects. *Ecotoxicology and Environmental Safety, 271*, 115992. DOI: 10.1016/j.ecoenv.2024.115992 PMID: 38262092

Jha, Y. (2023). Regulation of photosynthesis under stress. In *Improving Stress Resilience in Plants: Physiological and Biochemical Basis and Utilization in Breeding* (pp. 35–48). Elsevier. DOI: 10.1016/B978-0-443-18927-2.00002-9

Jia, A., Song, X., Li, S., Liu, Z., Liu, X., Han, Z., Gao, H., Gao, Q., Zha, Y., Liu, Y., Wu, X., & Wang, G. (2024). Biochar enhances soil hydrological function by improving the pore structure of saline soil. *Agricultural Water Management, 306*, 109170. DOI: 10.1016/j.agwat.2024.109170

Jingjing, W. A. N. G., & Xiaohua, Y. U. (2022). Wastewater irrigation and crop yield: A meta-analysis. *Journal of Integrative Agriculture, 21*(4), 1215–1224. DOI: 10.1016/S2095-3119(21)63853-4

Johnson, A., Singhal, N., & Hashmatt, M. (2011). Metal–plant interactions: toxicity and tolerance. In *Biomanagement of metal-contaminated soils* (pp. 29–63). Springer. DOI: 10.1007/978-94-007-1914-9_2

Juknys, R., Vitkauskaitė, G., Račaitė, M., & Venclovienė, J. (2012). The impacts of heavy metals on oxidative stress and growth of spring barley. *Central European Journal of Biology, 7*(2), 299–306.

Jyothsna, P. I., & Murthy, S. D. S. (2016). Review on role of salicylic acid in the alleviation of cadmium induced damage in plants. *Research Journal of Pharmaceutical, Biological and Chemical Sciences, 7*(4), 2746–2751.

Kacienė, G., Juknys, R., & Januškaitienė, I. (2017). The role of oxidative stress in spring barley cross-adaptation to different heavy metals. *Archives of Agronomy and Soil Science, 63*(8), 1037–1048. DOI: 10.1080/03650340.2016.1256474

Kakbra, R. F. (2024). Effect of seaweed, moringa leaf extract and biofertilizer on growth, yield and fruit quality of cucumber (*Cucumis sativus* L.) under greenhouse condition. arXiv. https://arxiv.org/abs/2403.17984

Kalai, T., Bouthour, D., Manai, J., Bettaieb Ben Kaab, L., & Gouia, H. (2016). Salicylic acid alleviates the toxicity of cadmium on seedling growth, amylases and phosphatases activity in germinating barley seeds. *Archives of Agronomy and Soil Science, 62*(6), 892–904. DOI: 10.1080/03650340.2015.1100295

Kalaivanan, D., & Ganeshamurthy, A. N. (2016). Mechanisms of heavy metal toxicity in plants. In *Abiotic stress physiology of horticultural crops* (pp. 85–102). Springer India., DOI: 10.1007/978-81-322-2725-0_5

Kalyabina, V. P., Esimbekova, E. N., Kopylova, K. V., & Kratasyuk, V. A. (2021). Pesticides: Formulants, distribution pathways and effects on human health—A review. *Toxicology Reports*, *8*, 1179–1192. DOI: 10.1016/j.toxrep.2021.06.004

Kang, M., Liu, Y., Weng, Y., Wang, H., & Bai, X. (2024). A critical review on the toxicity regulation and ecological risks of zinc oxide nanoparticles to plants. *Environmental Science. Nano*, *11*(1), 14–35. DOI: 10.1039/D3EN00630A

Kapoor, D., Bhardwaj, S., Gautam, S., Rattan, A., Bhardwaj, R., & Sharma, A. (2022). Brassinosteroids in plant nutrition and heavy metal tolerance. In *Brassinosteroids in plant developmental biology and stress tolerance* (pp. 217–235). Academic Press. DOI: 10.1016/B978-0-12-813227-2.00008-4

Karmous, I., Gammoudi, N., & Chaoui, A. (2023). Assessing the potential role of zinc oxide nanoparticles for mitigating cadmium toxicity in Capsicum annuum L. under in vitro conditions. *Journal of Plant Growth Regulation*, *42*(2), 719–734. DOI: 10.1007/s00344-022-10579-4

Karn, R., Ojha, N., Abbas, S., & Bhugra, S. (2021, June). A review on heavy metal contamination at mining sites and remedial techniques. [). IOP Publishing.]. *IOP Conference Series. Earth and Environmental Science*, *796*(1), 012013. DOI: 10.1088/1755-1315/796/1/012013

Kaur, R., Choudhary, D., Bali, S., Bandral, S. S., Singh, V., Ahmad, M. A., Rani, N., Singh, T. G., & Chandrasekaran, B. (2024). Pesticides: An alarming detrimental to health and environment. *The Science of the Total Environment*, *915*, 170113. DOI: 10.1016/j.scitotenv.2024.170113

Kaushik, S., Ranjan, A., Singh, A. K., & Sirhindi, G. (2024). Methyl jasmonate reduces cadmium toxicity by enhancing phenol and flavonoid metabolism and activating the antioxidant defense system in pigeon pea (*Cajanus cajan*). *Chemosphere*, *346*, 140681. DOI: 10.1016/j.chemosphere.2023.140681

Kaya, C., Akram, N. A., Sürücü, A., & Ashraf, M. (2019). Alleviating effect of nitric oxide on oxidative stress and antioxidant defence system in pepper (Capsicum annuum L.) plants exposed to cadmium and lead toxicity applied separately or in combination. *Scientia Horticulturae*, *255*, 52–60. DOI: 10.1016/j.scienta.2019.05.029

Kaya, C., Ugurlar, F., Ashraf, M., El-Sheikh, M. A., Bajguz, A., & Ahmad, P. (2022). The participation of nitric oxide in hydrogen sulphide-mediated chromium tolerance in pepper (Capsicum annuum L) plants by modulating subcellular distribution of chromium and the ascorbate-glutathione cycle. *Environmental Pollution*, *313*, 120229. DOI: 10.1016/j.envpol.2022.120229

Keramat, B., Kalantari, K. M., & Arvin, M. J. (2009). Effects of methyl jasmonate in regulating cadmium induced oxidative stress in soybean plant (*Glycine max* L.). *African Journal of Microbiological Research*, *3*(5), 240–244.

Keramat, B., Kalantari, K. M., & Arvin, M. J. (2010). Effects of methyl jasmonate treatment on alleviation of cadmium damages in soybean. *Journal of Plant Nutrition*, *33*(7), 1016–1025. DOI: 10.1080/01904161003728685

Keunen, E., Schellingen, K., Vangronsveld, J., & Cuypers, A. (2016). Ethylene and metal stress: Small molecule, big impact. *Frontiers in Plant Science*, *7*, 23. DOI: 10.3389/fpls.2016.00023

Khalid, M. F., Iqbal Khan, R., Jawaid, M. Z., Shafqat, W., Hussain, S., Ahmed, T., Rizwan, M., Ercisli, S., Pop, O. L., & Alina Marc, R. (2022). Nanoparticles: The plant saviour under abiotic stresses. *Nanomaterials (Basel, Switzerland)*, *12*(21), 3915. DOI: 10.3390/nano12213915 PMID: 36364690

Khan, E.A., Ahmed, H.M., Misra, M., Sharma, P. and Misra, A.N., 2022. Nitric Oxide Alleviates Photochemical Damage Induced by Cadmium Stress in Pea Seedlings. *Phyton (0031-9457)*, *91*(5).

Khan, M. I. R., Syeed, S., Nazar, R., & Anjum, N. A. (2012). An insight into the role of salicylic acid and jasmonic acid in salt stress tolerance. *Phytohormones and abiotic stress tolerance in plants*, pp.277-300.

Khan, M., Nawaz, N., Ali, I., Azam, M., Rizwan, M., Ahmad, P., & Ali, S. (2019). Regulation of photosynthesis under metal stress. In *Photosynthesis, productivity and environmental stress* (pp. 95–105). DOI: 10.1002/9781119501800.ch6

Khan, A. R., Mihoub, A., Jamal, A., Saeed, M. F., Soltani-Gerdefaramarzi, S., & Ahamad, M. I. (2025). Assessing Heavy Metal Contamination and Health Risks in Urban Street Dust: Insights from Pakistan's Industrial Growth. *Water, Air, and Soil Pollution*, *236*(11), 738. DOI: 10.1007/s11270-025-08359-3

Khan, A., Khan, S., Khan, M. A., Qamar, Z., & Waqas, M. (2015). The uptake and bioaccumulation of heavy metals by food plants, their effects on plants nutrients, and associated health risk: A review. *Environmental Science and Pollution Research International*, *22*(18), 13772–13799. DOI: 10.1007/s11356-015-4881-0

Khan, N., Bano, A. M., & Babar, A. (2020). Impacts of plant growth promoters and plant growth regulators on rainfed agriculture. *PLoS One*, *15*(4), e0231426. DOI: 10.1371/journal.pone.0231426

Khan, N., Ryu, K. Y., Choi, J. Y., Nho, E. Y., Habte, G., Choi, H., Kim, M. H., Park, K. S., & Kim, K. S. (2015). Determination of toxic heavy metals and speciation of arsenic in seaweeds from South Korea. *Food Chemistry*, *169*, 464–470. DOI: 10.1016/j.foodchem.2014.08.020

Khanna, P. (2011). Assessment of heavy metal contamination in different vegetables grown in and around urban areas. *Research Journal of Environmental Toxicology*, *5*(3), 162–179. DOI: 10.3923/rjet.2011.162.179

Khodanitska, O., Tkachuk, O., & Matviichuk, O. (2021). Physiological activity of plant growth stimulators. *The Scientific Heritage*, *58*(1), 36–38.

Khoulati, A., Ouahhoud, S., Taibi, M., Ezrari, S., Mamri, S., Merah, O., Hakkou, A., Addi, M., Maleb, A., & Saalaoui, E. (2025). Harnessing biostimulants for sustainable agriculture: Innovations, challenges, and future prospects. *Discover Agriculture, 3*(1), Article 56.

Khursheed, M. H. U. R., Shahbaz, M., Ramzan, T., Haider, A., Maqsood, M. F., Khan, A., Zulfiqar, U., Jamil, M., Hussain, S., Al-Ghamdi, A. A., & Rizwana, H. (2024). Enhancing wheat tolerance to cadmium stress through moringa leaf extract foliar application. *Scientifica*, *2024*(1), 2919557. DOI: 10.1155/2024/2919557

Kiran, B., Bharti, R., & Sharma, R. (2021). Effect of heavy metals: An overview. *Materials Today: Proceedings*, *51*, 880–885. DOI: 10.1016/j.matpr.2021.06.278

Kohli, S. K., Handa, N., Kaur, R., Kumar, V., Khanna, K., Bakshi, P., Singh, R., Arora, S., Kaur, R., & Bhardwaj, R. (2017). Role of salicylic acid in heavy metal stress tolerance: insight into underlying mechanism. In *Salicylic acid: a multifaceted hormone* (pp. 123–144). Springer Singapore. DOI: 10.1007/978-981-10-6068-7_7

Koleva, L., Umar, A., Yasin, N. A., Shah, A. A., Siddiqui, M. H., Alamri, S., Riaz, L., Raza, A., Javed, T., & Shabbir, Z. (2022). Iron oxide and silicon nanoparticles modulate mineral nutrient homeostasis and metabolism in cadmium-stressed Phaseolus vulgaris. *Frontiers in Plant Science*, *13*, 806781. DOI: 10.3389/fpls.2022.806781 PMID: 35386669

Körpe, D. A., & Aras, S. (2011). Evaluation of copper-induced stress on eggplant (Solanum melongena L.) seedlings at the molecular and population levels by use of various biomarkers. *Mutation Research. Genetic Toxicology and Environmental Mutagenesis*, *719*(1-2), 29–34. DOI: 10.1016/j.mrgentox.2010.10.003

Kour, J., Kohli, S. K., Khanna, K., Bakshi, P., Sharma, P., Singh, A. D., Ibrahim, M., Devi, K., Sharma, N., Ohri, P., Skalicky, M., Brestic, M., Bhardwaj, R., Landi, M., & Sharma, A. (2021). Brassinosteroid signaling, crosstalk and, physiological functions in plants under heavy metal stress. *Frontiers in Plant Science*, *12*, 608061. DOI: 10.3389/fpls.2021.608061

Kovács, V., Gondor, O. K., Szalai, G., Darkó, É., Majláth, I., Janda, T., & Pál, M. (2014). Synthesis and role of salicylic acid in wheat varieties with different levels of cadmium tolerance. *Journal of Hazardous Materials*, *280*, 12–19. DOI: 10.1016/j.jhazmat.2014.07.048

Krantev, A., Yordanova, R., Janda, T., Szalai, G., & Popova, L. (2008). Treatment with salicylic acid decreases the effect of cadmium on photosynthesis in maize plants. *Journal of Plant Physiology, 165*(9), 920–931. DOI: 10.1016/j.jplph.2006.11.014

Krasylenko, Y. A., Yemets, A. I., & Blume, Y. B. (2010). Functional role of nitric oxide in plants. *Russian Journal of Plant Physiology: a Comprehensive Russian Journal on Modern Phytophysiology, 57*(4), 451–461. DOI: 10.1134/S1021443710040011

Krishnani, K. K., Choudhary, K., Boddu, V. M., Moon, D. H., & Meng, X. (2021). Heavy metals biosorption mechanism of partially delignified products derived from mango (*Mangifera indica*) and guava (*Psidium guajava*) barks. *Environmental Science and Pollution Research International, 28*(25), 32891–32904. DOI: 10.1007/s11356-021-12874-1 PMID: 33638079

Kukde, S., Sarangi, B. K., & Purohit, H. (2019). Antioxidant role of nanoparticles for enhancing ecological performance of plant system. [). Elsevier.]. *Comprehensive Analytical Chemistry, 87*, 159–187. DOI: 10.1016/bs.coac.2019.09.004

Kumari, A., Gupta, A. K., Sharma, S., Jadon, V. S., Sharma, V., Chun, S. C., & Sivanesan, I. (2024). Nanoparticles as a tool for alleviating plant stress: Mechanisms, implications, and challenges. *Plants, 13*(11), 1528. DOI: 10.3390/plants13111528 PMID: 38891334

Kumar, M., Gogoi, A., Kumari, D., Borah, R., Das, P., Mazumder, P., & Tyagi, V. K. (2017). Review of perspective, problems, challenges, and future scenario of metal contamination in the urban environment. *Journal of Hazardous, Toxic and Radioactive Waste, 21*(4), 04017007. DOI: 10.1061/(ASCE)HZ.2153-5515.0000351

Kumar, S., Shah, S. H., Vimala, Y., Jatav, H. S., Ahmad, P., Chen, Y., & Siddique, K. H. (2022). Abscisic acid: Metabolism, transport, crosstalk with other plant growth regulators, and its role in heavy metal stress mitigation. *Frontiers in Plant Science, 13*, 972856. DOI: 10.3389/fpls.2022.972856

Kumar, S., & Trivedi, P. K. (2016). Heavy metal stress signalling in plants. In *Plant metal interaction: Emerging remediation techniques* (pp. 585–603). Elsevier. DOI: 10.1016/B978-0-12-803158-2.00025-4

Kumar, S., Wang, M., Fahad, S., Qayyum, A., Chen, Y., & Zhu, G. (2022). Chromium induces toxicity at different phenotypic, physiological, biochemical, and ultrastructural levels in sweet potato (Ipomoea batatas L.) plants. *International Journal of Molecular Sciences, 23*(21), 13496. DOI: 10.3390/ijms232113496

Küpper, H., & Andresen, E. (2016). Mechanisms of metal toxicity in plants. *Metallomics : Integrated Biometal Science, 8*(3), 269–285. DOI: 10.1039/C5MT00244C

Kuryntseva, P., Karamova, K., Galitskaya, P., Selivanovskaya, S., & Evtugyn, G. (2023). Biochar Functions in Soil Depending on Feedstock and Pyrolyzation Properties with Particular Emphasis on Biological Properties. *Agriculture, 13*(10), 2003. Advance online publication. DOI: 10.3390/agriculture13102003

Lanrewaju, A. A., & Fadiji, A. E. (2025). Microbial biostimulants: Boosting plant performance and stress tolerance. In *Microbial allies* (pp. 147–176). Springer. DOI: 10.1007/978-3-031-90530-8_8

Lan, X., Li, J., Chen, J., Liu, J., Cao, F., Liao, C., Zhang, Z., Gu, M., Wei, Y., Shen, F., Wei, X., Luo, X., & Zhang, X. (2023). Effects of foliar applications of Brassinolide and Selenium on the accumulation of Arsenic and Cadmium in rice grains and an assessment of their health risk. *International Journal of Phytoremediation, 25*(2), 161–171. DOI: 10.1080/15226514.2022.2066064

Laude, A. J., & Prior, I. A. (2004). Plasma membrane microdomains: Organization, function and trafficking. *Molecular Membrane Biology, 21*(3), 193–205. DOI: 10.1080/09687680410001700517 PMID: 15204627

Lee, M. S., Methela, N. J., Lee, G. H., & Mun, B. G. (2025). Nitric Oxide and Melatonin Cross Talk on Photosynthetic Machinery. *Molecules (Basel, Switzerland), 30*(10), 2148. DOI: 10.3390/molecules30102148

Lei, G. J., Sun, L., Sun, Y., Zhu, X. F., Li, G. X., & Zheng, S. J. (2020). Jasmonic acid alleviates cadmium toxicity in Arabidopsis via suppression of cadmium uptake and translocation. *Journal of Integrative Plant Biology*, *62*(2), 218–227. DOI: 10.1111/jipb.12801

Libutti, A., Trotta, V., & Rivelli, A. R. (2020). Biochar, vermicompost, and compost as soil organic amendments: Influence on Growth Parameters, Nitrate and Chlorophyll Content of Swiss Chard (Beta vulgaris L. var. cycla). *Agronomy (Basel)*, *10*(3), 346. DOI: 10.3390/agronomy10030346

Li, D., Zhou, C., Ma, J., Wu, Y., Kang, L., An, Q., Zhang, J., Deng, K., Li, J. Q., & Pan, C. (2021). Nanoselenium transformation and inhibition of cadmium accumulation by regulating the lignin biosynthetic pathway and plant hormone signal transduction in pepper plants. *Journal of Nanobiotechnology*, *19*(1), 316. DOI: 10.1186/s12951-021-01061-6

Li, Q., & Imran, . (2024). Mitigation strategies for heavy metal toxicity and its negative effects on soil and plants. *International Journal of Phytoremediation*, *26*(9), 1439–1452. DOI: 10.1080/15226514.2024.2327611 PMID: 38494751

Liu, C., Guo, J., Cui, Y., Lü, T., Zhang, X., & Shi, G. (2011). Effects of cadmium and salicylic acid on growth, spectral reflectance and photosynthesis of castor bean seedlings. *Plant and Soil*, *344*(1), 131–141. DOI: 10.1007/s11104-011-0733-y

Liu, H., & Timko, M. P. (2021). Jasmonic acid signaling and molecular crosstalk with other phytohormones. *International Journal of Molecular Sciences*, *22*(6), 2914. DOI: 10.3390/ijms22062914

Liu, J., Gai, L., & Zong, H. (2021). Foliage application of chitosan alleviates the adverse effects of cadmium stress in wheat seedlings (*Triticum aestivum* L.). *Plant Physiology and Biochemistry : PPB*, *164*, 115–121. DOI: 10.1016/j.plaphy.2021.04.038

Liu, X., Gong, D., Ke, Q., Yin, L., Wang, S., & Gao, T. (2023). Meta-analysis of the effect of nitric oxide application on heavy metal stress tolerance in plants. *Plants*, *12*(7), 1494. DOI: 10.3390/plants12071494

Liu, X., Yin, L., Deng, X., Gong, D., Du, S., Wang, S., & Zhang, Z. (2020). Combined application of silicon and nitric oxide jointly alleviated cadmium accumulation and toxicity in maize. *Journal of Hazardous Materials, 395*, 122679. DOI: 10.1016/j.jhazmat.2020.122679

Liu, Y., & Zhang, H. (2021). Reactive oxygen species and nitric oxide as mediators in plant hypersensitive response and stomatal closure. *Plant Signaling & Behavior, 16*(12), 1985860. DOI: 10.1080/15592324.2021.1985860

Li, Y., Sun, H., Liu, Z., Chu, Y., Huang, Y., & Bao, Q. (2024). Foliar jasmonic acid application reduces Cd and As accumulations in rice grains by regulating physiological, biochemical, and ROS scavenging attributes. *Environmental Technology & Innovation, 34*, 103596. DOI: 10.1016/j.eti.2024.103596

Loague, K., & Corwin, D. L. (2006). Point and nonpoint source pollution. In *Encyclopedia of hydrological sciences*.

Lucena, C., Romera, F. J., García, M. J., Alcántara, E., & Pérez-Vicente, R. (2015). Ethylene participates in the regulation of Fe deficiency responses in Strategy I plants and in rice. *Frontiers in Plant Science, 6*, 1056. DOI: 10.3389/fpls.2015.01056

Luo, Y., Wu, L., Liu, L., Han, C., & Li, Z. 2009. Heavy metal contamination and remediation in Asian agricultural land. In *MARCO symposium.*

Luo, P., Zhang, W., Xiao, D., Hu, J., Li, N., & Yang, J. (2025). Biochar-Based Fertilizers: Advancements, Applications, and Future Directions in Sustainable Agriculture—A Review. *Agronomy (Basel), 15*(5), 1104. DOI: 10.3390/agronomy15051104

Mahjoubi, Y., Emmerson, R., Kharbech, O., Massoud, M. B., Davey, P., & Djebali, W. (2025). Nitric oxide application improves photosynthetic light-response through enhancing photosystem II efficiency of bean plants (Phaseolus vulgaris L.) submitted to manganese stress. *Acta Physiologiae Plantarum, 47*(2), 24. DOI: 10.1007/s11738-025-03779-4

Maleki, M., Ghorbanpour, M., & Kariman, K. (2017). Physiological and antioxidative responses of medicinal plants exposed to heavy metals stress. *Plant Gene*, *11*, 247–254. DOI: 10.1016/j.plgene.2017.04.006

Małkowski, E., Sitko, K., Zieleźnik-Rusinowska, P., Gieroń, Ż., & Szopiński, M. (2019). Heavy metal toxicity: Physiological implications of metal toxicity in plants. In *Plant metallomics and functional omics: a system-wide perspective* (pp. 253–301). Springer International Publishing. DOI: 10.1007/978-3-030-19103-0_10

Mandal, A., Mundhe, D., Sonkamble, V., Wagh, N., & Lakkakula, J. (2022). Impact of heavy metal contamination on soil environment and advances in its revitalization strategies. In *Development in Waste Water Treatment Research and Processes* (pp. 215–241). Elsevier. DOI: 10.1016/B978-0-323-85584-6.00014-5

Manna, I., Nath, S., Mandal, P., & Paul, S. (2025). Effect of cerium oxide and silicon dioxide nanoparticles in combating heavy metal stress in plants. *The Nucleus : an International Journal of Cytology and Allied Topics*. Advance online publication. DOI: 10.1007/s13237-025-00582-x

Mansoor, S., Ali, A., Kour, N., Bornhorst, J., AlHarbi, K., Rinklebe, J., Abd El Moneim, D., Ahmad, P., & Chung, Y. S. (2023). Heavy metal induced oxidative stress mitigation and ROS scavenging in plants. *Plants*, *12*(16), 3003. DOI: 10.3390/plants12163003

Manzoor, M., Korolev, K. and Ahmad, R., (2025). Zinc oxide nanoparticles: abiotic stress tolerance in fruit crops focusing on sustainable production. Phyton (0031-9457), 94(5).

Manzoor, M., Korolev, K. and Ahmad, R., 2025. Zinc Oxide Nanoparticles: Abiotic Stress Tolerance in Fruit Crops Focusing on Sustainable Production. *Phyton (0031-9457), 94*(5).

Manzoor, H., Mehwish, , Bukhat, S., Rasul, S., Rehmani, M. I. A., Noreen, S., Athar, H.-R., Zafar, Z. U., Skalicky, M., Soufan, W., Brestic, M., Habib-ur-Rahman, M., Ogbaga, C. C., & EL Sabagh, A. (2022). Methyl jasmonate alleviated the adverse effects of cadmium stress in pea (Pisum sativum L.): A nexus of photosystem II activity and dynamics of redox balance. *Frontiers in Plant Science*, *13*, 860664. DOI: 10.3389/fpls.2022.860664

Manzoor, M., Korolev, K., & Ahmad, R. (2025). Zinc oxide nanoparticles: Abiotic stress tolerance in fruit crops focusing on sustainable production. *Phyton*, *94*(5), 1401–1418. DOI: 10.32604/phyton.2025.063930

Manzoor, N., Ali, L., Al-Huqail, A. A., Alghanem, S. M. S., Al-Haithloul, H. A. S., Abbas, T., Chen, G., Huan, L., Liu, Y., & Wang, G. (2023). Comparative efficacy of silicon and iron oxide nanoparticles towards improving the plant growth and mitigating arsenic toxicity in wheat (Triticum aestivum L.). *Ecotoxicology and Environmental Safety*, *264*, 115382. DOI: 10.1016/j.ecoenv.2023.115382 PMID: 37619453

Masarovičová, E., & Kráľová, K. (2012). *Plant-heavy metal interaction: phytoremediation, biofortification and nanoparticles. Advances in selected plant physiology aspects*. InTech.

Massoud, B. M., Sakouhi, L., & Chaoui, A. (2019). Effect of plant growth regulators, calcium and citric acid on copper toxicity in pea seedlings. *Journal of Plant Nutrition*, *42*(10), 1230–1242. DOI: 10.1080/01904167.2019.1609506

Mata-Pérez, C., Sánchez-Vicente, I., Arteaga, N., Gómez-Jiménez, S., Fuentes-Terrón, A., Oulebsir, C. S., Calvo-Polanco, M., Oliver, C., & Lorenzo, Ó. (2023). Functions of nitric oxide-mediated post-translational modifications under abiotic stress. *Frontiers in Plant Science*, *14*, 1158184. DOI: 10.3389/fpls.2023.1158184

Mediouni, C., Benzarti, O., Tray, B., Ghorbel, M. H., & Jemal, F. (2006). Cadmium and copper toxicity for tomato seedlings. *Agronomy for Sustainable Development*, *26*(4), 227–232. DOI: 10.1051/agro:2006008

Medvedkina, Y. A., & Khodochenko, A. V. (2020, October). Renewable energy and their impact on environmental pollution in the context of globalization. In 2020 International Multi-Conference on Industrial Engineering and Modern Technologies (FarEastCon) (pp. 1-4). IEEE.

Mehdizadeh, L., Moghaddam, M., Ganjeali, A., & Mahmoodi Sourestani, M. (2024). Changes in growth, essential oil composition and biochemical traits of peppermint in response to coapplication of zinc and methyl jasmonate in soilless culture. *Journal of Plant Nutrition*, *47*(16), 2732–2752. DOI: 10.1080/01904167.2024.2369063

Melebary, S. J. (2023). Heavy metal toxicity and remediation in human and agricultural systems: A review. *Advances in Animal and Veterinary Sciences*, *11*(4), 679–694. DOI: 10.17582/journal.aavs/2023/11.4.679.694

Mendoza-Soto, A. B., Sánchez, F., & Hernández, G. (2012). MicroRNAs as regulators in plant metal toxicity response. *Frontiers in Plant Science*, *3*, 105. DOI: 10.3389/fpls.2012.00105 PMID: 22661980

Meng, Y., Jing, H., Huang, J., Shen, R., & Zhu, X. (2022). The role of nitric oxide signaling in plant responses to cadmium stress. *International Journal of Molecular Sciences*, *23*(13), 6901. DOI: 10.3390/ijms23136901

Merwad, A. R. M. (2018). Using *Moringa oleifera* extract as biostimulant enhancing the growth, yield and nutrients accumulation of pea plants. *Journal of Plant Nutrition*, *41*(4), 425–431. DOI: 10.1080/01904167.2017.1384012

Mi, B., Liu, F., Xie, L., Zhou, H., Wu, F., & Dai, X. (2019). Evaluation of the uptake capacities of heavy metals in Chinese cabbage. *Ecotoxicology and Environmental Safety*, *171*, 511–517. DOI: 10.1016/j.ecoenv.2019.01.022 PMID: 30641311

Mihoub, A., Mesnoua, M., Touzout, N., Zeguerrou, R., Siabdallah, N., Benchikh, C., Benaoune, S., Jamal, A., Ronga, D., & Černý, J. (2024). Mitigation of Detrimental Effects of Salinity on Sweet Pepper through Biochar-Based Fertilizers Derived from Date Palm Wastes. *Phyton*, *93*(11), 2993–3011. DOI: 10.32604/phyton.2024.057536

Mir, M. A., John, R., Alyemeni, M. N., Alam, P., & Ahmad, P. (2018). Jasmonic acid ameliorates alkaline stress by improving growth performance, ascorbate glutathione cycle and glyoxylase system in maize seedlings. *Scientific Reports*, *8*(1), 2831. DOI: 10.1038/s41598-018-21097-3

Mishra, R. K. (2023). Fresh water availability and its global challenge. *British Journal of Multidisciplinary and Advanced Studies*, *4*(3), 1–78. DOI: 10.37745/bjmas.2022.0208

Mishra, V., Singh, P., Tripathi, D. K., Corpas, F. J., & Singh, V. P. (2021). Nitric oxide and hydrogen sulfide: An indispensable combination for plant functioning. *Trends in Plant Science*, *26*(12), 1270–1285. DOI: 10.1016/j.tplants.2021.07.016

Mittova, V., Guy, M., Tal, M., & Volokita, M. (2004). Salinity up-regulates the antioxidative system in root mitochondria and peroxisomes of the wild salt-tolerant tomato species *Lycopersicon pennellii. Journal of Experimental Botany*, *55*(399), 1105–1113. DOI: 10.1093/jxb/erh113 PMID: 15047761

Mohamed, H. I., & Latif, H. H. (2017). Improvement of drought tolerance of soybean plants by using methyl jasmonate. *Physiology and Molecular Biology of Plants : an International Journal of Functional Plant Biology*, *23*(3), 545–556. DOI: 10.1007/s12298-017-0451-x

Mondal, S., Pramanik, K., Ghosh, S. K., Pal, P., Ghosh, P. K., Ghosh, A., & Maiti, T. K. (2022). Molecular insight into arsenic uptake, transport, phytotoxicity, and defense responses in plants: A critical review. *Planta*, *255*(4), 87. DOI: 10.1007/s00425-022-03869-4

Mourato, M. P., Moreira, I. N., Leitão, I., Pinto, F. R., Sales, J. R., & Louro Martins, L. (2015). Effect of heavy metals in plants of the genus *Brassica. International Journal of Molecular Sciences*, *16*(8), 17975–17998. DOI: 10.3390/ijms160817975 PMID: 26247945

Mousavi, S. M. (2022). Silicon and nano-silicon mediated heavy metal stress tolerance in plants. In *Silicon and nano-silicon in environmental stress management and crop quality improvement* (pp. 181–191). Academic Press. DOI: 10.1016/B978-0-323-91225-9.00012-1

Mousavi, S. R., Niknejad, Y., Fallah, H., & Tari, D. B. (2020). Methyl jasmonate alleviates arsenic toxicity in rice. *Plant Cell Reports*, *39*(8), 1041–1060. DOI: 10.1007/s00299-020-02547-7

Muhammad, H.M.D., Naz, S., Bibi, Z., Ahmed, M.A., Stanciu, A.S. & Ahmad, R. (2025). Nanoparticles and Phytohormonal Synergy in Plants: Sustainable Agriculture Approach. *Phyton (0031-9457), 94*(9).

Muhammad, H.M.D., Naz, S., Bibi, Z., Ahmed, M.A., Stanciu, A.S. and Ahmad, R., 2025. Nanoparticles and Phytohormonal Synergy in Plants: Sustainable Agriculture Approach. *Phyton (0031-9457), 94*(9).

Muhammad, H. M. D., Naz, S., Bibi, Z., Ahmed, M. A., Stanciu, A. S., & Ahmad, R. (2025). Nanoparticles and phytohormonal synergy in plants: Sustainable agriculture approach. *Phyton*, *94*(9), 2631–2648. DOI: 10.32604/phyton.2025.069474

Muhammad, H. M. D., Naz, S., Lal, M. K., Tiwari, R. K., Ahmad, R., Nawaz, M. A., Das, R., & Altaf, M. A. (2024). Melatonin in business with abiotic stresses in vegetable crops. *Scientia Horticulturae*, *324*, 112594. DOI: 10.1016/j.scienta.2023.112594

Munyaneza, V., Zhang, W., Haider, S., Ren, L., Ali, A., Kant, S., & Ding, G. (2025). Mitigating Metal Toxicity in Plants Using Nanoparticles: Mechanisms and Implications for Sustainable Agriculture. *Journal of Agricultural and Food Chemistry*, *73*(28), 17374–17386. DOI: 10.1021/acs.jafc.5c04693 PMID: 40601963

Mushtaq, R., Shafiq, M., Batool, A., Din, M. I., & Sami, A. (2024). Investigate the impact of zinc oxide nanoparticles under lead toxicity on Chilli (Capsicum annuum L). *Bulletin of Biological and Allied Sciences Research*, *2024*(1), 90–90. DOI: 10.54112/bbasr.v2024i1.90

Mushtaq, T., Bano, A., & Ullah, A. (2025). Effects of Rhizospheric Microbes, Growth Regulators, and Biochar in Modulating Antioxidant Machinery of Plants Under Stress. *Journal of Plant Growth Regulation*, *44*(5), 1846–1867. DOI: 10.1007/s00344-024-11548-9

Mustafa, G., & Komatsu, S. (2016). Toxicity of heavy metals and metal-containing nanoparticles on plants. *Biochimica et Biophysica Acta (BBA)-. Biochimica et Biophysica Acta. Proteins and Proteomics, 1864*(8), 932–944. DOI: 10.1016/j.bbapap.2016.02.020

Naaz, H., Rawat, K., Saffeullah, P., & Umar, S. (2023). Silica nanoparticles synthesis and applications in agriculture for plant fertilization and protection: A review. *Environmental Chemistry Letters, 21*(1), 539–559. DOI: 10.1007/s10311-022-01515-9

Nagajyoti, P. C., Lee, K. D., & Sreekanth, T. V. M. (2010). Heavy metals, occurrence and toxicity for plants: A review. *Environmental Chemistry Letters, 8*(3), 199–216. DOI: 10.1007/s10311-010-0297-8

Nahar, K., Rhaman, M. S., Parvin, K., Bardhan, K., Marques, D. N., García-Caparrós, P., & Hasanuzzaman, M. (2022). Arsenic-induced oxidative stress and antioxidant defense in plants. *Stresses, 2*(2), 179–209. DOI: 10.3390/stresses2020013

Nawaz, M., Saleem, M. H., Khalid, M. R., Ali, B., & Fahad, S. (2024). Nitric oxide reduces cadmium uptake in wheat (Triticum aestivum L.) by modulating growth, mineral uptake, yield attributes, and antioxidant profile. *Environmental Science and Pollution Research International, 31*(6), 9844–9856. DOI: 10.1007/s11356-024-31875-4

Nazir, F., Jahan, B., Iqbal, N., Rajurkar, A. B., Siddiqui, M. H., & Khan, M. I. R. (2023). Methyl jasmonate influences ethylene formation, defense systems, nutrient homeostasis and carbohydrate metabolism to alleviate arsenic-induced stress in rice (*Oryza sativa*). *Plant Physiology and Biochemistry : PPB, 202*, 107990. DOI: 10.1016/j.plaphy.2023.107990

Naz, S., Anjum, M. A., Sadiq, B., Ahmad, R., Altaf, M. A., El-Sheikh, M. A., & Shakoor, A. (2022). Purification of sewage wastewater through sand column filter for lessening heavy metal accumulation in lettuce, carrot, and cauliflower. *Water (Basel), 14*(22), 3770. DOI: 10.3390/w14223770

Neyshabouri, F. A., Ghotbi-Ravandi, A. A., Shariatmadari, Z., & Tohidfar, M. (2024). Cadmium toxicity promotes hormonal imbalance and induces the expression of genes involved in systemic resistances in barley. *Biometals : An International Journal on the Role of Metal Ions in Biology, Biochemistry, and Medicine*, *37*(5), 1147–1160. DOI: 10.1007/s10534-024-00597-y

Nguyen, T. H., Goossens, A., & Lacchini, E. (2022). Jasmonate: A hormone of primary importance for plant metabolism. *Current Opinion in Plant Biology*, *67*, 102197. DOI: 10.1016/j.pbi.2022.102197

Nguyen, T. Q., Sesin, V., Kisiala, A., & Emery, R. N. (2021). Phytohormonal roles in plant responses to heavy metal stress: Implications for using macrophytes in phytoremediation of aquatic ecosystems. *Environmental Toxicology and Chemistry*, *40*(1), 7–22. DOI: 10.1002/etc.4909

Niu, L., Zhao, H., Tang, Y., Zhu, B., Zhao, Y., Wang, Q., & Yu, J. (2025). Methyl jasmonate was involved in hydrogen sulfide-alleviated cadmium stress in cucumber plants through ROS homeostasis and chlorophyll metabolism. *International Journal of Molecular Sciences*, *26*(2), 475. DOI: 10.3390/ijms26020475

Noctor, G., & Foyer, C. H. (1998). Ascorbate and glutathione: Keeping active oxygen under control. *Annual Review of Plant Physiology and Plant Molecular Biology*, *49*(1), 249–279. DOI: 10.1146/annurev.arplant.49.1.249 PMID: 15012235

Noman, A., & Aqeel, M. (2017). miRNA-based heavy metal homeostasis and plant growth. *Environmental Science and Pollution Research International*, *24*(11), 10068–10082. DOI: 10.1007/s11356-017-8593-5

Nowicka, B. (2022). Heavy metal–induced stress in eukaryotic algae—Mechanisms of heavy metal toxicity and tolerance with particular emphasis on oxidative stress in exposed cells and the role of antioxidant response. *Environmental Science and Pollution Research International*, *29*(12), 16860–16911. DOI: 10.1007/s11356-021-18419-w

Nwankwo, C. E. I., Okeke, E. S., Umeoguaju, F. U., Ejeromedoghene, O., Adedipe, D. T., & Ezeorba, T. P. C. (2025). Addressing emerging contaminants in agriculture affecting plant–soil interaction: A review on bio-based and nano-enhanced strategies for soil health and global food security (GFS). *Discover Toxicology*, *2*(1), 1–30. DOI: 10.1007/s44339-025-00018-w

Ochoa, L., Shrivastava, M., Srivastava, S., Cota-Ruiz, K., Zhao, L., White, J. C., Hernandez-Viezcas, J. A., & Gardea-Torresdey, J. L. (2025). Nanomaterials for managing abiotic and biotic stress in the soil–plant system for sustainable agriculture. *Environmental Science. Nano*, *12*(2), 1037–1058. DOI: 10.1039/D4EN00789A

Ogugua, U. V., Kanu, S. A., & Ntushelo, K. (2022). Gibberellic acid improves growth and reduces heavy metal accumulation: A case study in tomato (*Solanum lycopersicum* L.) seedlings exposed to acid mine water. *Heliyon*, *8*(12), e12399. DOI: 10.1016/j.heliyon.2022.e12399

Okant, M., & Kaya, C. (2019). The role of endogenous nitric oxide in melatonin-improved tolerance to lead toxicity in maize plants. *Environmental Science and Pollution Research International*, *26*(12), 11864–11874. DOI: 10.1007/s11356-019-04517-3

Oladimeji, T. E., Oyedemi, M., Emetere, M. E., Agboola, O., Adeoye, J. B., & Odunlami, O. A. (2024). Review on the impact of heavy metals from industrial wastewater effluent and removal technologies. *Heliyon*, *10*(23), e40370. DOI: 10.1016/j.heliyon.2024.e40370

Oliveira, D. J. B., Lavres, J., Kopittke, P. M., Chaney, R. L., Harris, H. H., Erskine, P. D., Howard, D. L., dos Reis, A. R., & van der Ent, A. (2024). Unravelling the fate of foliar-applied nickel in soybean: A comprehensive investigation. *Plant and Soil*, *502*(1), 537–556. DOI: 10.1007/s11104-024-06567-0

Ondrasek, G., Rengel, Z., & Romic, D. (2018). Humic acids decrease uptake and distribution of trace metals, but not the growth of radish exposed to cadmium toxicity. *Ecotoxicology and Environmental Safety*, *151*, 55–61. DOI: 10.1016/j.ecoenv.2017.12.055

Özkay, F., Kıran, S., Kuşvuran, Ş., & Ellialtıoğlu, Ş. Ş. (2016). The effects of humic acid applications on heavy metal stress in lettuce. *Turkish Journal of Agriculture - Food Science and Technology*, *4*(6), 431–437.

Palansooriya, K. N., Wong, J. T. F., Hashimoto, Y., Huang, L., Rinklebe, J., Chang, S. X., Bolan, N., Wang, H., & Ok, Y. S. (2019). Response of microbial communities to biochar-amended soils: A critical review. *Biochar*, *1*(1), 3–22. DOI: 10.1007/s42773-019-00009-2

Pál, M., Janda, T., & Szalai, G. (2018). Interactions between plant hormones and thiol-related heavy metal chelators. *Plant Growth Regulation*, *85*(2), 173–185. DOI: 10.1007/s10725-018-0391-7

Pant, N., Kumar, G., Upadhyay, A. D., Patel, D. K., Gupta, Y. K., & Chaturvedi, P. K. (2014). Reproductive toxicity of lead, cadmium, and phthalate exposure in men. *Environmental Science and Pollution Research International*, *21*(18), 11066–11074. DOI: 10.1007/s11356-014-2986-5

Pan, W., Lu, Q., Xu, Q. R., Zhang, R. R., Li, H. Y., Yang, Y. H., Liu, H. J., & Du, S. T. (2019). Abscisic acid-generating bacteria can reduce Cd concentration in pakchoi grown in Cd-contaminated soil. *Ecotoxicology and Environmental Safety*, *177*, 100–107. DOI: 10.1016/j.ecoenv.2019.04.010

Pardo-Hernández, M., López-Delacalle, M., Martí-Guillen, J. M., Martínez-Lorente, S. E., & Rivero, R. M. (2021). ROS and NO phytomelatonin-induced signaling mechanisms under metal toxicity in plants: A review. *Antioxidants*, *10*(5), 775. DOI: 10.3390/antiox10050775

Pascual, L. S., Mittler, R., Sinha, R., Peláez-Vico, M. Á., López-Climent, M. F., Vives-Peris, V., Gómez-Cadenas, A., & Zandalinas, S. I. (2023). Jasmonic acid is required for tomato acclimation to multifactorial stress combination. *Environmental and Experimental Botany*, *213*, 105425. DOI: 10.1016/j.envexpbot.2023.105425

Pasković, I., Popović, L., Pongrac, P., Polić Pasković, M., Kos, T., Jovanov, P., & Franić, M. (2024). Protein hydrolysates—Production, effects on plant metabolism, and use in agriculture. *Horticulturae*, *10*(10), 1041. DOI: 10.3390/horticulturae10101041

Patel, M., Surti, M. M., Ashraf, S. A., & Adnan, M. (2021). Physiological and molecular responses to heavy metal stresses in plants. In *Harsh environment and plant resilience* (pp. 171–202). DOI: 10.1007/978-3-030-65912-7_8

Paudel, I., Bar-Tal, A., Levy, G. J., Rotbart, N., Ephrath, J. E., & Cohen, S. (2018). Treated wastewater irrigation: Soil variables and grapefruit tree performance. *Agricultural Water Management*, *204*, 126–137. DOI: 10.1016/j.agwat.2018.04.006

Peñarrubia, L., Romero, P., Carrió-Seguí, A., Andrés-Bordería, A., Moreno, J., & Sanz, A. (2015). Temporal aspects of copper homeostasis and its crosstalk with hormones. *Frontiers in Plant Science*, *6*, 255. DOI: 10.3389/fpls.2015.00255

Perelomov, L. V., Sarkar, B., Sizova, O. I., Chilachava, K. B., Shvikin, A. Y., Perelomova, I. V., & Atroshchenko, Y. M. (2018). Zinc and lead detoxifying abilities of humic substances relevant to environmental bacterial species. *Ecotoxicology and Environmental Safety*, *151*, 178–183. DOI: 10.1016/j.ecoenv.2018.01.018

Per, T. S., Khan, N. A., Masood, A., & Fatma, M. (2016). Methyl jasmonate alleviates cadmium-induced photosynthetic damages through increased S-assimilation and glutathione production in mustard. *Frontiers in Plant Science*, *7*, 1933. DOI: 10.3389/fpls.2016.01933

Piotrowska-Niczyporuk, A., Bajguz, A., Zambrzycka-Szelewa, E., & Bralska, M. (2018). Exogenously applied auxins and cytokinins ameliorate lead toxicity by inducing antioxidant defence system in green alga Acutodesmus obliquus. *Plant Physiology and Biochemistry : PPB*, *132*, 535–546. DOI: 10.1016/j.plaphy.2018.09.038

Piotto, F. A., Carvalho, M. E. A., Souza, L. A., Rabêlo, F. H. S., Franco, M. R., Batagin-Piotto, K. D., & Azevedo, R. A. (2018). Estimating tomato tolerance to heavy metal toxicity: Cadmium as study case. *Environmental Science and Pollution Research International*, *25*(27), 27535–27544. DOI: 10.1007/s11356-018-2778-4 PMID: 30051291

Piršelová, B., Galuščáková, Ľ., Lengyelová, L., Kubová, V., Jandová, V., & Hegrová, J. (2022). Assessment of the hormetic effect of arsenic on growth and physiology of two cultivars of maize (*Zea mays* L.). *Plants*, *11*(24), 3433. DOI: 10.3390/plants11243433 PMID: 36559544

Polle, A., & Schützendübel, A. (2003). Heavy metal signalling in plants: Linking cellular and organismic responses. In *Plant responses to abiotic stress* (pp. 187–215). Springer. DOI: 10.1007/978-3-540-39402-0_8

Popova, L. P., Maslenkova, L. T., Ivanova, A., & Stoinova, Z. (2011). Role of salicylic acid in alleviating heavy metal stress. In *Environmental adaptations and stress tolerance of plants in the era of climate change* (pp. 447–466). Springer New York.

Popova, L., Maslenkova, L., Yordanova, R., Krantev, A., Szalai, G., & Janda, T. (2008). Salicylic acid protects photosynthesis against cadmium toxicity in pea plants. *General and Applied Plant Physiology*, *34*(3-4), 133–148.

Prasad, R. (2022). Cytokinin and its key role to enrich the plant nutrients and growth under adverse conditions-an update. *Frontiers in Genetics*, *13*, 883924. DOI: 10.3389/fgene.2022.883924

Praveen, A., Khan, E., Ngiimei, D. S., Perwez, M., Sardar, M., & Gupta, M. (2018). Iron oxide nanoparticles as nano-adsorbents: A possible way to reduce arsenic phytotoxicity in Indian mustard plant (Brassica juncea L.). *Journal of Plant Growth Regulation*, *37*(2), 612–624. DOI: 10.1007/s00344-017-9760-0

Pretty, J. (2008). Agricultural sustainability: Concepts, principles and evidence. *Philosophical Transactions of the Royal Society of London. Series B, Biological Sciences*, *363*(1491), 447–465. DOI: 10.1098/rstb.2007.2163 PMID: 17652074

Qadir, M., Wichelns, D., Raschid-Sally, L., McCornick, P. G., Drechsel, P., Bahri, A., & Minhas, P. S. (2010). The challenges of wastewater irrigation in developing countries. *Agricultural Water Management*, *97*(4), 561–568. DOI: 10.1016/j.agwat.2008.11.004

Qin, C., Shen, J., & Ahanger, M. A. (2022). Supplementation of nitric oxide and spermidine alleviates the nickel stress-induced damage to growth, chlorophyll metabolism, and photosynthesis by upregulating ascorbate–glutathione and glyoxalase cycle functioning in tomato. *Frontiers in Plant Science*, *13*, 1039480. DOI: 10.3389/fpls.2022.1039480

Raffa, C. M., & Chiampo, F. (2021). Bioremediation of agricultural soils polluted with pesticides: A review. *Bioengineering (Basel, Switzerland)*, *8*(7), 92. DOI: 10.3390/bioengineering8070092

Rahman, M. H., Roy, B., Chowdhury, G. M., Hasan, A., & Saimun, M. S. R. (2022). Medicinal plant sources and traditional healthcare practices of forest-dependent communities in and around Chunati Wildlife Sanctuary in southeastern Bangladesh. *Environmental Sustainability*, *5*(2), 207–241. DOI: 10.1007/s42398-022-00230-z PMID: 37521586

Rahman, S. U., Li, Y., Hussain, S., Hussain, B., Khan, W. U. D., Riaz, L., Ashraf, M. N., Khaliq, M. A., Du, Z., & Cheng, H. (2023). Role of phytohormones in heavy metal tolerance in plants: A review. *Ecological Indicators*, *146*, 109844. DOI: 10.1016/j.ecolind.2022.109844

Rahman, S., Ahmad, I., & Nafees, M. (2023). Mitigation of heavy metal stress in maize (Zea mays L.) through application of silicon nanoparticles. *Biocatalysis and Agricultural Biotechnology*, *50*, 102757. DOI: 10.1016/j.bcab.2023.102757

Rai, G. K., Bhat, B. A., Mushtaq, M., Tariq, L., Rai, P. K., Basu, U., Dar, A. A., Islam, S. T., Dar, T. U., & Bhat, J. A. (2021). Insights into decontamination of soils by phytoremediation: A detailed account on heavy metal toxicity and mitigation strategies. *Physiologia Plantarum*, *173*(1), 287–304. DOI: 10.1111/ppl.13433

Rai, P. K., Lee, S. S., Zhang, M., Tsang, Y. F., & Kim, K. H. (2019). Heavy metals in food crops: Health risks, fate, mechanisms, and management. *Environment International*, *125*, 365–385. DOI: 10.1016/j.envint.2019.01.067

Rajewska, I., Talarek, M., & Bajguz, A. (2016). Brassinosteroids and response of plants to heavy metals action. *Frontiers in Plant Science*, *7*, 629. DOI: 10.3389/fpls.2016.00629

Rajput, V. D., Minkina, T., Feizi, M., Kumari, A., Khan, M., Mandzhieva, S., Sushkova, S., El-Ramady, H., Verma, K. K., Singh, A., Hullebusch, E. D. V., Singh, R. K., Jatav, H. S., & Choudhary, R. (2021). Effects of silicon and silicon-based nanoparticles on rhizosphere microbiome, plant stress and growth. *Biology (Basel)*, *10*(8), 791. DOI: 10.3390/biology10080791 PMID: 34440021

Rana, M. S., Chen, H., Deng, S., Imran, M., Abdellah, Y. A. Y., Li, W., Lin, J., Li, J., & Wang, R. (2025). Biological insights into alleviating heavy metal toxicity through the simultaneous supply of biochar biofilters derived from the Mikania micrantha and molybdenum nanoparticles. *Biochar*, *7*(1), 23. Advance online publication. DOI: 10.1007/s42773-024-00414-2

Rani, A., Kumar, A., Lal, A., & Pant, M. (2014). Cellular mechanisms of cadmium-induced toxicity: A review. *International Journal of Environmental Health Research*, *24*(4), 378–399. DOI: 10.1080/09603123.2013.835032

Rani, S., Kumari, N., & Sharma, V. (2025). Zinc Oxide Nanoparticles Improve Photosynthesis by Modulating Antioxidant System and psb A Gene Expression Under Arsenic Stress in Different Cultivars of Vigna radiata. *BioNanoScience*, *15*(2), 242. DOI: 10.1007/s12668-025-01858-x

Ran, S., Zhang, K., Zhou, Y., Huang, W., & Zhong, F. (2023). Nickel Ions Enhanced the Adaptability of Tomato Seedling Roots to Low-Nitrogen Stress by Improving Their Antioxidant Capacity. *Horticulturae*, *9*(12), 1342. DOI: 10.3390/horticulturae9121342

Rao, M. J., Duan, M., Zhou, C., Jiao, J., Cheng, P., Yang, L., Wei, W., Shen, Q., Ji, P., Yang, Y., Conteh, O., Yan, D., Yuan, H., Rauf, A., Ai, J., & Zheng, B. (2025). Antioxidant defense system in plants: Reactive oxygen species production, signaling, and scavenging during abiotic stress-induced oxidative damage. *Horticulturae*, *11*(5), 477. DOI: 10.3390/horticulturae11050477

Rascio, N., & Navari-Izzo, F. (2011). Heavy metal hyperaccumulating plants: How and why do they do it? And what makes them so interesting? *Plant Science : An International Journal of Experimental Plant Biology*, *180*(2), 169–181. DOI: 10.1016/j.plantsci.2010.08.016 PMID: 21421358

Rasheed, A., Hassan, M. U., Fahad, S., Aamer, M., Batool, M., Ilyas, M., & Li, H. (2021). Heavy metals stress and plants defense responses. In *Sustainable soil and land management and climate change* (pp. 57–82). CRC Press.

Rasheed, R., Ashraf, M. A., Arshad, A., Iqbal, M., & Hussain, I. (2020). Interactive effects of chitosan and cadmium on growth, secondary metabolism, oxidative defense, and element uptake in pea (*Pisum sativum* L.). *Arabian Journal of Geosciences*, *13*(17), 847. DOI: 10.1007/s12517-020-05871-0

Rashid, A., Schutte, B. J., Ulery, A., Deyholos, M. K., Sanogo, S., Lehnhoff, E. A., & Beck, L. (2023). Heavy metal contamination in agricultural soil: Environmental pollutants affecting crop health. *Agronomy (Basel)*, *13*(6), 1521. DOI: 10.3390/agronomy13061521

Rastogi, A., Zivcak, M., Sytar, O., Kalaji, H. M., He, X., Mbarki, S., & Brestic, M. (2017). Impact of metal and metal oxide nanoparticles on plant: A critical review. *Frontiers in Chemistry*, *5*, 78. DOI: 10.3389/fchem.2017.00078 PMID: 29075626

Rathinapriya, P., Maharajan, T., Jothi, R., Prabakaran, M., Lee, I. B., Yi, P. H., & Jeong, S. T. (2024). Unlocking biochar impacts on abiotic stress dynamics: A systematic review of soil quality and crop improvement. *Frontiers in Plant Science*, *15*, 1479925. DOI: 10.3389/fpls.2024.1479925

Rathod, S. V., Saras, P., & Gondaliya, S. M. (2024). Environmental pollution: Threats and challenges for management. In *Eco-restoration of polluted environment* (pp. 1–34). CRC Press. DOI: 10.1201/9781003423393-1

Rawat, S., Pullagurala, V. L., Hernandez-Molina, M., Sun, Y., Niu, G., Hernandez-Viezcas, J. A., Peralta-Videa, J. R., & Gardea-Torresdey, J. L. (2018). Impacts of copper oxide nanoparticles on bell pepper (Capsicum annum L.) plants: A full life cycle study. *Environmental Science. Nano*, *5*(1), 83–95. DOI: 10.1039/C7EN00697G

Raza, A., Charagh, S., Zahid, Z., Mubarik, M. S., Javed, R., Siddiqui, M. H., & Hasanuzzaman, M. (2021). Jasmonic acid: A key frontier in conferring abiotic stress tolerance in plants. *Plant Cell Reports*, *40*(8), 1513–1541. DOI: 10.1007/s00299-020-02614-z

Raza, A., Hussain, S., Javed, R., Hafeez, M. B., & Hasanuzzaman, M. (2021). Antioxidant defense systems and remediation of metal toxicity in plants. In *Approaches to the remediation of inorganic pollutants* (pp. 91–124). Springer Singapore. DOI: 10.1007/978-981-15-6221-1_6

Raza, A., Salehi, H., Rahman, M. A., Zahid, Z., Madadkar Haghjou, M., Najafi-Kakavand, S., Charagh, S., Osman, H. S., Albaqami, M., Zhuang, Y., Siddique, K. H., & Zhuang, W. (2022). Plant hormones and neurotransmitter interactions mediate antioxidant defenses under induced oxidative stress in plants. *Frontiers in Plant Science*, *13*, 961872. DOI: 10.3389/fpls.2022.961872

Razzaq, S., Zhou, B., Zia-Ur-rehman, M., Maqsood, M. A., Hussain, S., Bakhsh, G., Zhang, Z., Yang, Q., & Altaf, A. R. (2022). Cadmium Stabilization and Redox Transformation Mechanism in Maize Using Nanoscale Zerovalent-Iron-Enriched Biochar in Cadmium-Contaminated Soil. *Plants*, *11*(8), 1074. DOI: 10.3390/plants11081074

Rehman, M., Saeed, M. S., Fan, X., Salam, A., Munir, R., Yasin, M. U., Khan, A. R., Muhammad, S., Ali, B., Ali, I., Khan, J., & Gan, Y. (2023). The multifaceted role of jasmonic acid in plant stress mitigation: An overview. *Plants*, *12*(23), 3982. DOI: 10.3390/plants12233982

Ren, J., Ren, X., Deng, Z., Zhang, H., Wang, J., Zhang, C., Lu, F., & Shi, J. (2025). Ecological effects of biochar in heavy metal-contaminated soils from multidimensional perspective: Using meta-analysis. *Bioresource Technology*, *432*, 132695. DOI: 10.1016/j.biortech.2025.132695

Repkina, N., Murzina, S. A., Voronin, V. P., & Kaznina, N. (2023). Does methyl jasmonate effectively protect plants under heavy metal contamination? Fatty acid content in wheat leaves exposed to cadmium with or without exogenous methyl jasmonate application. *Biomolecules*, *13*(4), 582. DOI: 10.3390/biom13040582

Riyazuddin, R., Nisha, N., Ejaz, B., Khan, M. I. R., Kumar, M., Ramteke, P. W., & Gupta, R. (2021). A comprehensive review on the heavy metal toxicity and sequestration in plants. *Biomolecules*, *12*(1), 43. DOI: 10.3390/biom12010043 PMID: 35053191

Rizvi, A., Zaidi, A., Ameen, F., Ahmed, B., AlKahtani, M. D., & Khan, M. S. (2020). Heavy metal induced stress on wheat: Phytotoxicity and microbiological management. *RSC Advances*, *10*(63), 38379–38403. DOI: 10.1039/D0RA05610C

Rong, Q., Zhong, K., Huang, H., Li, C., Zhang, C., & Nong, X. (2020). Humic acid reduces the available cadmium, copper, lead, and zinc in soil and their uptake by tobacco. *Applied Sciences (Basel, Switzerland)*, *10*(3), 1077. DOI: 10.3390/app10031077

Rostami, S., & Azhdarpoor, A. (2019). The application of plant growth regulators to improve phytoremediation of contaminated soils: A review. *Chemosphere*, *220*, 818–827. DOI: 10.1016/j.chemosphere.2018.12.203

Roychowdhury, R., Khan, M. H., & Choudhury, S. (2019). Physiological and molecular responses for metalloid stress in rice—A Comprehensive Overview. *Advances in rice research for abiotic stress tolerance*, pp.341-369.

Roy, T. K., Islam, M. S., Mahiddin, N. A., Hossain, S. A., Biswas, T., Antu, U. B., Serity, S. A., Miti, J. F., Akter, S., Roy, S., Biswas, A., Mazrin, M., & Ismail, Z. (2025). Application of nanoparticles (NPs) to ameliorate abiotic stress in economically important crop species: A potential review. *Journal of Crop Health*, *77*(1), 19. DOI: 10.1007/s10343-024-01069-6 PMID: 41649711

Rroji, M., Figurek, A., Viggiano, D., Capasso, G., & Spasovski, G. (2022). Phosphate in the context of cognitive impairment and other neurological disorders occurrence in chronic kidney disease. *International Journal of Molecular Sciences*, *23*(13), 7362. DOI: 10.3390/ijms23137362 PMID: 35806367

Ruciñiska-Sobkowiak, R. (2010). Oxidative stress in plants exposed to heavy metals. *Postępy Biochemii*, *56*(2), 191–200.

Rucińska-Sobkowiak, R. (2016). Water relations in plants subjected to heavy metal stresses. *Acta Physiologiae Plantarum*, *38*(11), 257. DOI: 10.1007/s11738-016-2277-5

Sah, S. K., Mhetre, A. G., Kumari, S., Gadi, Y., Prajapat, M., Kashyap, M., Adhikari, S., & Kumar, A. (2025). Emerging techniques of soil remediation: Nanotechnology and bioremediation approaches. *International Journal of Advanced Biochemistry Research*, *9*(1), 488–504. DOI: 10.33545/26174693.2025.v9.i1f.3560

Saini, S., Kaur, N., & Pati, P. K. (2021). Phytohormones: Key players in the modulation of heavy metal stress tolerance in plants. *Ecotoxicology and Environmental Safety*, *223*, 112578. DOI: 10.1016/j.ecoenv.2021.112578

Salazar, M., González, E., Casaretto, J. A., Casacuberta, J. M., & Ruiz-Lara, S. (2007). The promoter of the TLC1. 1 retrotransposon from Solanum chilense is activated by multiple stress-related signaling molecules. *Plant Cell Reports*, *26*(10), 1861–1868. DOI: 10.1007/s00299-007-0375-y

Salifu, M., John, M. A., Abubakar, M., Bankole, I. A., Ajayi, N. D., & Amusan, O. (2024). Phytoremediation strategies for heavy metal contamination: A Review on sustainable approach for environmental restoration. *Journal of Environmental Protection*, *15*(4), 450–474. DOI: 10.4236/jep.2024.154026

Sami, A., Shah, F. A., Abdullah, M., Zhou, X., Yan, Y., Zhu, Z., & Zhou, K. (2020). Melatonin mitigates cadmium and aluminium toxicity through modulation of antioxidant potential in *Brassica napus* L. *Plant Biology*, *22*(4), 679–690. DOI: 10.1111/plb.13093

Sami, F., Faizan, M., Faraz, A., Siddiqui, H., Yusuf, M., & Hayat, S. (2018). Nitric oxide-mediated integrative alterations in plant metabolism to confer abiotic stress tolerance, NO crosstalk with phytohormones and NO-mediated post translational modifications in modulating diverse plant stress. *Nitric Oxide : Biology and Chemistry*, *73*, 22–38. DOI: 10.1016/j.niox.2017.12.005

Sardar, K., Ali, S., Hameed, S., Afzal, S., Fatima, S., Shakoor, M. B., Bharwana, S. A., & Tauqeer, H. M. (2013). Heavy metals contamination and what are the impacts on living organisms. *Greener Journal of Environmental Management and Public Safety*, *2*(4), 172–179. DOI: 10.15580/GJEMPS.2013.4.060413652

Sarim, M., Jan, T., Khattak, S. A., Mihoub, A., Jamal, A., Saeed, M. F., Soltani-Gerdefaramarzi, S., Tariq, S. R., Fernández, M. P., Mancinelli, R., & Radicetti, E. (2022). Assessment of the Ecological and Health Risks of Potentially Toxic Metals in Agricultural Soils from the Drosh-Shishi Valley, Pakistan. *Land (Basel)*, *11*(10), 1663. DOI: 10.3390/land11101663

Saxena, I., & Shekhawat, G. S. (2013). Nitric oxide (NO) in alleviation of heavy metal induced phytotoxicity and its role in protein nitration. *Nitric Oxide : Biology and Chemistry*, *32*, 13–20. DOI: 10.1016/j.niox.2013.03.004

Schutzendubel, A., & Polle, A. (2002). Plant responses to abiotic stresses: Heavy metal-induced oxidative stress and protection by mycorrhization. *Journal of Experimental Botany*, *53*(372), 1351–1365. DOI: 10.1093/jexbot/53.372.1351

Seifried, H. E., Anderson, D. E., Fisher, E. I., & Milner, J. A. (2007). A review of the interaction among dietary antioxidants and reactive oxygen species. *The Journal of Nutritional Biochemistry*, *18*(9), 567–579. DOI: 10.1016/j.jnutbio.2006.10.007

Seneviratne, M., Rajakaruna, N., Rizwan, M., Madawala, H., Ok, Y. S., & Vithanage, M. (2019). Heavy metal-induced oxidative stress on seed germination and seedling development: A critical review. *Environmental Geochemistry and Health*, *41*(4), 1813–1831. DOI: 10.1007/s10653-017-0005-8

Shaikh, A., Jamla, M., Sharma, L., Palawat, B., Pathak, S., Shekhawat, G. S., Tripathi, M., & Kumar, V. (2025). Multifaceted roles of biogenic nanoparticles for conferring heavy metal tolerance in plants. *Journal of Crop Health*, *77*(3), 82. DOI: 10.1007/s10343-025-01147-3

Shar, A. G., Zhang, L., Lu, A., Ahmad, M., Saqib, M., Hussain, S., Zulfiqar, U., Wang, P., Zhang, L., & Rahimi, M. (2025). Unlocking biochar's potential: Innovative strategies for sustainable remediation of heavy metal stress in tobacco plants. *Scientifica*, *2025*(1), 6302968. DOI: 10.1155/sci5/6302968

Sharma, S., Singh, V. K., Kumar, A., & Mallubhotla, S. 2019. Effect of nanoparticles on oxidative damage and antioxidant defense system in plants. *Molecular plant abiotic stress: Biology and biotechnology*, pp.315-333.

Sharma, A., Dubey, S., Singh, K., Mittal, R., Quille, P., & Rajauria, G. (2025). Innovative processing and industrial applications of seaweed. *Phycology*, *5*(1), 10. DOI: 10.3390/phycology5010010

Sharma, A., Shahzad, B., Kumar, V., Kohli, S. K., Sidhu, G. P. S., Bali, A. S., Handa, N., Kapoor, D., Bhardwaj, R., & Zheng, B. (2019). Phytohormones regulate accumulation of osmolytes under abiotic stress. *Biomolecules*, *9*(7), 285. DOI: 10.3390/biom9070285

Sharma, A., Sidhu, G. P. S., Araniti, F., Bali, A. S., Shahzad, B., Tripathi, D. K., Brestic, M., Skalicky, M., & Landi, M. (2020). The role of salicylic acid in plants exposed to heavy metals. *Molecules (Basel, Switzerland)*, *25*(3), 540. DOI: 10.3390/molecules25030540

Sharma, A., Soares, C., Sousa, B., Martins, M., Kumar, V., Shahzad, B., Sidhu, G. P., Bali, A. S., Asgher, M., Bhardwaj, R., Thukral, A. K., Fidalgo, F., & Zheng, B. (2020). Nitric oxide-mediated regulation of oxidative stress in plants under metal stress: A review on molecular and biochemical aspects. *Physiologia Plantarum*, *168*(2), 318–344. DOI: 10.1111/ppl.13004

Sharma, M. M. M., Kapoor, D., Loyal, A., Kumar, R., Sharma, P., & Husen, A. (2024). *Plant Response to Silver Nanoparticles in Terms of Growth, Development, Production, and Protection: An Overview. Plant Response to Silver Nanoparticles: Plant Growth.* Development, Production, and Protection.

Sharma, R. K., & Agrawal, M. (2005). Biological effects of heavy metals: An overview. *Journal of Environmental Biology*, *26*(2), 301–313.

Sharma, S. S., & Dietz, K. J. (2009). The relationship between metal toxicity and cellular redox imbalance. *Trends in Plant Science, 14*(1), 43–50. DOI: 10.1016/j.tplants.2008.10.007

Sharma, S., Raja, V., Bhat, A. H., Kumar, N., Alsahli, A. A., & Ahmad, P. (2025). Innovative strategies for alleviating chromium toxicity in tomato plants using melatonin functionalized zinc oxide nanoparticles. *Scientia Horticulturae, 341*, 113930. DOI: 10.1016/j.scienta.2024.113930

Shi, J., Li, J., Pan, Y., Zhao, M., Zhang, R., Xue, Y., & Liu, Y. (2024). The physiological response mechanism of peanut leaves under Al stress. *Plants, 13*(12), 1606. DOI: 10.3390/plants13121606

Shikha, D., Jakhar, P., & Satbhai, S. B. (2023). Role of jasmonate signaling in the regulation of plant responses to nutrient deficiency. *Journal of Experimental Botany, 74*(4), 1221–1243. DOI: 10.1093/jxb/erac387

Shivappa, S., Amritha, K. P., Nayak, S., Chandrashekar, H. K., Thorat, S. A., Kaniyassery, A., Govender, N., Thiruvengadam, M., & Muthusamy, A. (2025). Integration of physio-biochemical, biological and molecular approaches to improve heavy metal tolerance in plants. *3 Biotech, 15*(4), 1–28.

Shoaib, A., & Javaid, A. 2021. Oxidative stress in plants exposed to heavy metals. In *Organic Solutes, Oxidative Stress, and Antioxidant Enzymes Under Abiotic Stressors* (pp. 133-152). CRC Press. DOI: 10.1201/9781003022879-7

Shukla, P. K., Misra, P., Maurice, N., & Ramteke, P. W. (2019). Heavy metal toxicity and possible functional aspects of microbial diversity in heavy metal-contaminated sites. In *Microbial Genomics in Sustainable Agroecosystems* (Vol. 2, pp. 255–317). Springer Singapore. DOI: 10.1007/978-981-32-9860-6_15

Shu, X., Yin, L., Zhang, Q., & Wang, W. (2012). Effect of Pb toxicity on leaf growth, antioxidant enzyme activities, and photosynthesis in cuttings and seedlings of Jatropha curcas L. *Environmental Science and Pollution Research International, 19*(3), 893–902. DOI: 10.1007/s11356-011-0625-y

Siddiqi, K. S., & Husen, A. (2019). Plant response to jasmonates: Current developments and their role in changing environment. *Bulletin of the National Research Center*, *43*(1), 1. DOI: 10.1186/s42269-019-0195-6

Siddiqui, M. H., Al-Whaibi, M. H., Ali, H. M., Sakran, A. M., Basalah, M. O., & AlKhaishany, M. Y. (2013). Mitigation of nickel stress by the exogenous application of salicylic acid and nitric oxide in wheat. *Australian Journal of Crop Science*, *7*(11), 1780–1788.

Simek, J., Tuma, J., Dohnal, V., Musil, K., & Ducaiová, Z. (2016). Salicylic acid and phenolic compounds under cadmium stress in cucumber plants (Cucumis sativus L.). *Acta Physiologiae Plantarum*, *38*(7), 172. DOI: 10.1007/s11738-016-2192-9

Singh, A. D., Sharma, P., Kohli, S. K., Kumar, P., Singh, R., Arora, P., Sharma, P., Kaur, R., Sharma, A., & Bhardwaj, R. (2020). Role of plant growth regulators (PGRs) in mitigation of heavy metal phytotoxicity in plants. In *Cellular and molecular phytotoxicity of heavy metals* (pp. 263–304). Springer International Publishing. DOI: 10.1007/978-3-030-45975-8_14

Singhal, R. K., Jatav, H. S., Aftab, T., Pandey, S., Mishra, U. N., Chauhan, J., Chand, S., Indu, , Saha, D., Dadarwal, B. K., Chandra, K., Khan, M. A., Rajput, V. D., Minkina, T., Narayana, E. S., Sharma, M. K., & Ahmed, S. (2021). Roles of nitric oxide in conferring multiple abiotic stress tolerance in plants and crosstalk with other plant growth regulators. *Journal of Plant Growth Regulation*, *40*(6), 2303–2328. DOI: 10.1007/s00344-021-10446-8

Singhal, R. K., Kumar, M., Bose, B., Mondal, S., Srivastava, S., Dhankher, O. P., & Tripathi, R. D. (2023). Heavy metal (loid) s phytotoxicity in crops and its mitigation through seed priming technology. *International Journal of Phytoremediation*, *25*(2), 187–206. DOI: 10.1080/15226514.2022.2068502

Singh, B. N., Mishra, R. N., Agarwal, P. K., Goswami, M., Nair, S., Sopory, S. K., & Reddy, M. K. (2004). A pea chloroplast translation elongation factor that is regulated by abiotic factors. *Biochemical and Biophysical Research Communications*, *320*(2), 523–530. DOI: 10.1016/j.bbrc.2004.05.192

Singh, H. P., Kaur, S., Batish, D. R., Sharma, V. P., Sharma, N., & Kohli, R. K. (2009). Nitric oxide alleviates arsenic toxicity by reducing oxidative damage in the roots of Oryza sativa (rice). *Nitric Oxide : Biology and Chemistry*, *20*(4), 289–297. DOI: 10.1016/j.niox.2009.02.004

Singh, I., & Shah, K. (2014). Exogenous application of methyl jasmonate lowers the effect of cadmium-induced oxidative injury in rice seedlings. *Phytochemistry*, *108*, 57–66. DOI: 10.1016/j.phytochem.2014.09.007

Singh, R., Choudhary, P., Kumar, S., & Daima, H. K. (2024). Mechanistic approaches for crosstalk between nanomaterials and plants: Plant immunomodulation, defense mechanisms, stress resilience, toxicity, and perspectives. *Environmental Science. Nano*, *11*(6), 2324–2351. DOI: 10.1039/D4EN00053F

Sinha, P., Shukla, A. K., & Sharma, Y. K. (2015). Amelioration of heavy-metal toxicity in cauliflower by application of salicylic acid. *Communications in Soil Science and Plant Analysis*, *46*(10), 1309–1319. DOI: 10.1080/00103624.2015.1033543

Sinha, S., Basak, A., & Mondol, M. S. A. (2025). Understanding heavy metal accumulation in crops: Sources, plant responses, tolerance mechanisms, and environmental effects. *Journal of Environmental Science and Health. Part C, Toxicology and Carcinogenesis*, *43*(3), 269–294. DOI: 10.1080/26896583.2025.2521207

Sirhindi, G., Mir, M. A., Abd-Allah, E. F., Ahmad, P., & Gucel, S. (2016). Jasmonic acid modulates the physio-biochemical attributes, antioxidant enzyme activity, and gene expression in Glycine max under nickel toxicity. *Frontiers in Plant Science*, *7*, 591. DOI: 10.3389/fpls.2016.00591

Širić, I., Eid, E. M., El-Morsy, M. H., Osman, H. E., Adelodun, B., Abou Fayssal, S., Mioč, B., Goala, M., Singh, J., Bachheti, A., Arya, A. K., Choi, K. S., Kumar, V., & Kumar, P. (2022). Health risk assessment of hazardous heavy metals in two varieties of mango fruit (Mangifera indica L. var. Dasheri and Langra). *Horticulturae*, *8*(9), 832. DOI: 10.3390/horticulturae8090832

Sofy, M. R., Seleiman, M. F., Alhammad, B. A., Alharbi, B. M., & Mohamed, H. I. (2020). Minimizing adverse effects of pb on maize plants by combined treatment with jasmonic, salicylic acids and proline. *Agronomy (Basel)*, *10*(5), 699. DOI: 10.3390/agronomy10050699

Sohn, S. I., Pandian, S., Rakkammal, K., Largia, M. J. V., Thamilarasan, S. K., Balaji, S., Zoclanclounon, Y. A. B., Shilpha, J., & Ramesh, M. (2022). Jasmonates in plant growth and development and elicitation of secondary metabolites: An updated overview. *Frontiers in Plant Science*, *13*, 942789. DOI: 10.3389/fpls.2022.942789

Soliman, M., Alhaithloul, H. A., Hakeem, K. R., Alharbi, B. M., El-Esawi, M., & Elkelish, A. (2019). Exogenous nitric oxide mitigates nickel-induced oxidative damage in eggplant by upregulating antioxidants, osmolyte metabolism, and glyoxalase systems. *Plants*, *8*(12), 562. DOI: 10.3390/plants8120562

Song, C., Bibi, Z., Ali, S., & Ahmad, R. (2026). Silicon mitigates aluminum toxicity in lettuce by improving photosynthetic pigments and antioxidant defense system. *Biometals*, 1–14. DOI: 10.1007/s10534-026-00819-5 PMID: 42050262

Song, W. Y., Yang, H. C., Shao, H. B., Zheng, A. Z., & Brestic, M. (2014). The alleviative effects of salicylic acid on the activities of catalase and superoxide dismutase in malting barley (Hordeum uhulgare L.) seedling leaves stressed by heavy metals. *Clean (Weinheim)*, *42*(1), 88–97. DOI: 10.1002/clen.201200310

Soni, S., Jha, A. B., Dubey, R. S., & Sharma, P. (2024). Mitigating cadmium accumulation and toxicity in plants: The promising role of nanoparticles. *The Science of the Total Environment*, *912*, 168826. DOI: 10.1016/j.scitotenv.2023.168826 PMID: 38042185

Sood, M. (2023). Jasmonates: "The master switch" for regulation of developmental and stress responses in plants. *Journal of Plant Growth Regulation*, *42*(8), 5247–5265. DOI: 10.1007/s00344-023-11047-3

Souri, Z., Khanna, K., Karimi, N., & Ahmad, P. (2021). Silicon and plants: Current knowledge and future prospects. *Journal of Plant Growth Regulation, 40*(3), 906–925. DOI: 10.1007/s00344-020-10172-7

Sreelatha, L., Ambili, A. L., Sreedevi, S. C., & Achuthavarier, D. (2025). Metallothioneins: An unraveling insight into remediation strategies of plant defense mechanisms. *Environmental Science and Pollution Research International, 32*(2), 405–427. DOI: 10.1007/s11356-024-35790-6

Srikanth, H., Kumari, K., & Meena, R. (2022). Contamination of groundwater with heavy metals and its impact on soil and plants—A review. *Ecology. Environmental Conservation, 28*, S331–S335.

Srivastava, A. K., Kumari, S., Singh, R. P., Khan, M., Mishra, P., & Xie, X. (2025). Harnessing the interplay of protein posttranslational modifications: Enhancing plant resilience to heavy metal toxicity. *Microbiological Research*, •••, 128112.

Srivastava, V., Sarkar, A., Singh, S., Singh, P., De Araujo, A. S., & Singh, R. P. (2017). Agroecological responses of heavy metal pollution with special emphasis on soil health and plant performances. *Frontiers in Environmental Science, 5*, 64. DOI: 10.3389/fenvs.2017.00064

Steffens, B. (2014). The role of ethylene and ROS in salinity, heavy metal, and flooding responses in rice. *Frontiers in Plant Science, 5*, 685. DOI: 10.3389/fpls.2014.00685

Subin, M. P., & Francis, S. (2013). Phytotoxic effects of cadmium on seed germination, early seedling growth and antioxidant enzyme activities in *Cucurbita maxima* Duchesne. *International Research Journal of Biological Sciences, 2*(9), 40–47.

Suman, S. N., Pradhan, B., Kumar, V., Singh, H., Nath, D., Kamat, D. N., & Kumar, M. (2025). Exploring the role of biostimulants in sustainable agriculture. *Current Science, 128*(1), 25–30. DOI: 10.18520/cs/v128/i1/25-30

Sun, L., Wang, R., Ju, Q., Xing, M., Li, R., Li, W., Li, W., Wang, W., Deng, Y., & Xu, J. (2023). Mitigation mechanism of zinc oxide nanoparticles on cadmium toxicity in tomato. *Frontiers in Plant Science*, *14*, 1162372. DOI: 10.3389/fpls.2023.1162372

Sun, Z., Su, C., Yun, J., Jiang, Q., Wang, L., Wang, Y., Cao, D., Zhao, F., Zhao, Q., Zhang, M., Zhou, B., Zhang, L., Kong, F., Liu, B., Tong, Y., & Li, X. (2019). Genetic improvement of the shoot architecture and yield in soya bean plants via the manipulation of GmmiR156b. *Plant Biotechnology Journal*, *17*(1), 50–62. DOI: 10.1111/pbi.12946 PMID: 29729214

Syed, R.S., Dhriti Kapoor, D.K. and Bhat, A.A., 2018. Heavy metal toxicity in plants: a review.

Sytar, O., Kumar, A., Latowski, D., Kuczynska, P., Strzałka, K., & Prasad, M. N. V. (2013). Heavy metal-induced oxidative damage, defense reactions, and detoxification mechanisms in plants. *Acta Physiologiae Plantarum*, *35*(4), 985–999. DOI: 10.1007/s11738-012-1169-6

Sytar, O., Kumari, P., Yadav, S., Brestic, M., & Rastogi, A. (2019). Phytohormone priming: Regulator for heavy metal stress in plants. *Journal of Plant Growth Regulation*, *38*(2), 739–752. DOI: 10.1007/s00344-018-9886-8

Tahir, R., Afzal, F., Jamil, H., Razzaq, M., & Khan, M. S. (2024). Physiological impacts of pesticidal contamination: Challenge to sustainable agriculture and biodegradation methods. *Advances in Agriculture and Animal Science*, *40*(1), 24–37. DOI: 10.47432/2024.40.1.4

Tak, H. I., Ahmad, F., & Babalola, O. O. (2012). Advances in the application of plant growth-promoting rhizobacteria in phytoremediation of heavy metals. *Reviews of Environmental Contamination and Toxicology*, *223*, 33–52.

Tamás, L., Mistrík, I., Alemayehu, A., Zelinová, V., Bočová, B., & Huttová, J. (2015). Salicylic acid alleviates cadmium-induced stress responses through the inhibition of Cd-induced auxin-mediated reactive oxygen species production in barley root tips. *Journal of Plant Physiology*, *173*, 1–8. DOI: 10.1016/j.jplph.2014.08.018

Tang, Y., Wang, L., Xie, Y., Yu, X., Li, H., Lin, L., Liao, M. A., Wang, Z., Sun, G., Wang, X., Liang, D., Xia, H., & Tu, L. (2020). Effects of exogenous abscisic acid on the growth and cadmium accumulation of lettuce under cadmium-stress conditions. *International Journal of Environmental Analytical Chemistry, 100*(6), 720–731. DOI: 10.1080/03067319.2019.1639686

Tapia, G., Verdugo, I., Yanez, M., Ahumada, I., Theoduloz, C., Cordero, C., Poblete, F., González, E., & Ruiz-Lara, S. (2005). Involvement of ethylene in stress-induced expression of the TLC1. 1 retrotransposon from Lycopersicon chilense Dun. *Plant Physiology, 138*(4), 2075–2086. DOI: 10.1104/pp.105.059766

Tassi, E., Pouget, J., Petruzzelli, G., & Barbafieri, M. (2008). The effects of exogenous plant growth regulators in the phytoextraction of heavy metals. *Chemosphere, 71*(1), 66–73. DOI: 10.1016/j.chemosphere.2007.10.027

Terrón-Camero, L. C., Peláez-Vico, M. Á., Del-Val, C., Sandalio, L. M., & Romero-Puertas, M. C. (2019). Role of nitric oxide in plant responses to heavy metal stress: Exogenous application versus endogenous production. *Journal of Experimental Botany, 70*(17), 4477–4488. DOI: 10.1093/jxb/erz184

Thakur, S., Walia, S., Thakur, B., Ghosh, A., & Kumar, R. (2025). Unleashing the potential of seaweed biostimulants—A comparative evaluation for enhancing saffron (*Crocus sativus* L.) yield with different corm sizes in the Western Himalayas. *BMC Plant Biology, 25*(1), 946. DOI: 10.1186/s12870-025-06955-3

Thao, N. P., Khan, M. I. R., Thu, N. B. A., Hoang, X. L. T., Asgher, M., Khan, N. A., & Tran, L. S. P. (2015). Role of ethylene and its cross talk with other signaling molecules in plant responses to heavy metal stress. *Plant Physiology, 169*(1), 73–84. DOI: 10.1104/pp.15.00663

Tighe-Neira, R., Carmora, E., Recio, G., Nunes-Nesi, A., Reyes-Diaz, M., Alberdi, M., Rengel, Z., & Inostroza-Blancheteau, C. (2018). Metallic nanoparticles influence the structure and function of the photosynthetic apparatus in plants. *Plant Physiology and Biochemistry : PPB, 130*, 408–417. DOI: 10.1016/j.plaphy.2018.07.024 PMID: 30064097

Touzout, N., Bouchibane, M., Tahraoui, H., Mihoub, A., Zhang, J., Amrane, A., Ahmad, I., Danish, S., Alahmadi, T. A., & Ansari, M. J. (2024c). Silicon-mediated resilience: Unveiling the protective role against combined cypermethrin and hymexazol phytotoxicity in tomato seedlings. *Journal of Environmental Management*, *369*, 122370. DOI: 10.1016/j.jenvman.2024.122370

Touzout, N., Mehallah, H., Moralent, R., Moulay, M., & Nemmiche, S. (2021b). Phytotoxic evaluation of neonicotinoid imidacloprid and cadmium alone and in combination on tomato (Solanum lycopersicum L.). *Ecotoxicology (London, England)*, *30*(6), 1126–1137. DOI: 10.1007/s10646-021-02421-6

Touzout, N., Mehallah, H., Moralent, R., Nemmiche, S., & Benkhelifa, M. (2021). Co-contamination of deltamethrin and cadmium induce oxidative stress in tomato plants (*Solanum lycopersicum* L.). *Acta Physiologiae Plantarum*, *43*(6), 91. DOI: 10.1007/s11738-021-03261-x

Touzout, N., Mihoub, A., Ahmad, I., Jamal, A., & Danish, S. (2024a). Deciphering the role of nitric oxide in mitigation of systemic fungicide induced growth inhibition and oxidative damage in wheat. *Chemosphere*, *364*, 143046. DOI: 10.1016/j.chemosphere.2024.143046

Touzout, N., Mihoub, A., Boukheddimi, M., Moualdia, A. O., Ahmad, I., Jamal, A., Danish, S., Alarfaj, A. A., Alharbi, S. A., & Javed Ansari, M. (2024b). Nitric oxide application alleviates fungicide and ampicillin co-exposure induced phytotoxicity by regulating antioxidant defense, detoxification system, and secondary metabolism in wheat seedlings. *Journal of Environmental Management*, *372*, 123337. DOI: 10.1016/j.jenvman.2024.123337

Touzout, N., Mihoub, A., Krache, F., Ali, E., & Ahmad, R. (2025). Biochar Improves Growth Traits, Photosynthetic Pigments, Antioxidant Enzymes, and Metabolites Accumulation of Wheat (Triticum aestivum L) Under Nickel Toxicity. *Journal of Soil Science and Plant Nutrition*, *25*(3), 1–16. DOI: 10.1007/s42729-025-02482-0

Tripathi, D. K., Singh, S., Singh, V. P., Prasad, S. M., Chauhan, D. K., & Dubey, N. K. (2016). Silicon nanoparticles more efficiently alleviate arsenate toxicity than silicon in maize cultiver and hybrid differing in arsenate tolerance. *Frontiers in Environmental Science*, *4*, 46. DOI: 10.3389/fenvs.2016.00046

Tu, C. Y., Wang, H. Y., Jiang, Q., Gao, Y. Q., Shen, R. F., & Zhu, X. F. (2025). Jasmonic acid enhances aluminum tolerance of rice by regulating detoxification genes. *Plant Physiology and Biochemistry : PPB*, *228*, 110257. DOI: 10.1016/j.plaphy.2025.110257

Ul Haq, I., Ahmed, U., Ali Khan, I., Rehman, A., Saeed, M. F., Alkahtani, J., Elshikh, M. S., Černý, J., & Jamal, A. (2025). Nano zinc-enriched biochar improves spinach growth and soil nutrient dynamics in degraded soils. *Journal of Plant Nutrition*, *48*(15), 2632–2647. DOI: 10.1080/01904167.2025.2487122

Ullah, A., Hazrat, A., Khan, B. A., Saqib, S., & Ullah, F. (2025). Mitigating lead stress in barley using gibberellic acid (GA3): Effects on morpho-physiological and biochemical parameters. *Journal of Plant Growth Regulation*, *44*(7), 1–12. DOI: 10.1007/s00344-025-11654-2

Ullah, N., Ur Rehman, M., Ahmad, B., Ali, I., Younas, M., Aslam, M. S., Rahman, A. U., Taheri, E., Fatehizadeh, A., & Rezakazemi, M. (2022). Assessment of heavy metals accumulation in agricultural soil, vegetables and associated health risks. *PLoS One*, *17*(6), e0267719. DOI: 10.1371/journal.pone.0267719

Ulloa-Inostroza, E. M., Alberdi, M., Ivanov, A. G., & Reyes-Díaz, M. (2019). Protective effect of methyl jasmonate on photosynthetic performance and its association with antioxidants in contrasting aluminum-resistant blueberry cultivars exposed to aluminum. *Journal of Soil Science and Plant Nutrition*, *19*(1), 203–216. DOI: 10.1007/s42729-019-0006-z

Uruç Parlak, K. (2016). Effect of nickel on growth and biochemical characteristics of wheat (*Triticum aestivum* L.) seedlings. *NJAS Wageningen Journal of Life Sciences*, *76*(1), 1–5. DOI: 10.1016/j.njas.2012.07.001

Van Nguyen, Q., Tran, H. M. T., Nguyen, H. T., Van Hoang, H., & Nguyen, K. M. (2025). The Effect of pH on Nitrogen and Phosphorus Recovery Efficiency From Wastewater Using Biochar Derived From Corncob. *Water, Air, and Soil Pollution*, *236*(11), 1–13. DOI: 10.1007/s11270-025-08377-1

Verma, G., Srivastava, D., Narayan, S., Shirke, P. A., & Chakrabarty, D. (2020). Exogenous application of methyl jasmonate alleviates arsenic toxicity by modulating its uptake and translocation in rice (*Oryza sativa* L.). *Ecotoxicology and Environmental Safety*, *201*, 110735. DOI: 10.1016/j.ecoenv.2020.110735

Verma, N., & Prasad, S. M. (2021). Interplay of hydrogen peroxide and nitric oxide: Systemic regulation of photosynthetic performance and nitrogen metabolism in cadmium challenged cyanobacteria. *Physiology and Molecular Biology of Plants : an International Journal of Functional Plant Biology*, *27*(10), 2181–2199. DOI: 10.1007/s12298-021-01083-2

Verma, N., Tiwari, S., Singh, V. P., & Prasad, S. M. (2020). Nitric oxide in plants: An ancient molecule with new tasks. *Plant Growth Regulation*, *90*(1), 1–13. DOI: 10.1007/s10725-019-00543-w

Verma, R., Singh, A., Khare, S., & Kumar, P. (2024). Alleviation of environmental stresses in crop plants by nanoparticles: Recent advances and future perspectives. *Journal of Plant Biochemistry and Biotechnology*, *34*(3), 1–24.

Viehweger, K. (2014). How plants cope with heavy metals. *Botanical Studies*, *55*(1), 35. DOI: 10.1186/1999-3110-55-35 PMID: 28510963

Vojodi Mehrabani, L. (2019). The effects of methanol and ethanol foliar application under salinity stress on some physiological characteristics of Pelargonium graveolens L. *Journal of Plant Physiology and Breeding*, *9*(1), 63–73.

Wang, X., Cui, J., Li, X., Wang, G., Zheng, G., Li, Y., Li, X., Ma, L., & Wang, L. (2022). Comparative study on growth, physiological and biotic stress resistance of cucumber (*Cucumis sativus* L.) by exogenous application of jasmonic acid and methyl jasmonate under greenhouse conditions. ... (Complete journal details as needed)

Wang, F., Tan, H., Huang, L., Cai, C., Ding, Y., Bao, H., Chen, Z., & Zhu, C. (2021). Application of exogenous salicylic acid reduces Cd toxicity and Cd accumulation in rice. *Ecotoxicology and Environmental Safety*, *207*, 111198. DOI: 10.1016/j.ecoenv.2020.111198

Wang, J., Song, L., Gong, X., Xu, J., & Li, M. (2020). Functions of jasmonic acid in plant regulation and response to abiotic stress. *International Journal of Molecular Sciences*, *21*(4), 1446. DOI: 10.3390/ijms21041446

Wang, M., Chen, L., Wu, L., Zhang, L., Xie, H., & Shen, Z. (2022). Review of nonpoint source pollution models: Current status and future direction. *Water (Basel)*, *14*(20), 3217. DOI: 10.3390/w14203217

Wang, R., Huang, J., Liang, A., Wang, Y., Mur, L. A. J., Wang, M., & Guo, S. (2020). Zinc and copper enhance cucumber tolerance to fusaric acid by mediating its distribution and toxicity and modifying the antioxidant system. *International Journal of Molecular Sciences*, *21*(9), 3370. DOI: 10.3390/ijms21093370

Wang, Y., Mostafa, S., Zeng, W., & Jin, B. (2021). Function and mechanism of jasmonic acid in plant responses to abiotic and biotic stresses. *International Journal of Molecular Sciences*, *22*(16), 8568. DOI: 10.3390/ijms22168568

Wani, A. B., Chadar, H., Wani, A. H., Singh, S., & Upadhyay, N. (2017). Salicylic acid to decrease plant stress. *Environmental Chemistry Letters*, *15*(1), 101–123. DOI: 10.1007/s10311-016-0584-0

Wani, Z. A., Ahmad, Z., Asgher, M., Bhat, J. A., Sharma, M., Kumar, A., Sharma, V., Kumar, A., Pant, S., Lukatkin, A. S., & Anjum, N. A. (2023). Phytoremediation of potentially toxic elements: Role, status and concerns. *Plants*, *12*(3), 429. DOI: 10.3390/plants12030429

Wasternack, C., & Hause, B. (2013). Jasmonates: Biosynthesis, perception, signal transduction and action in plant stress response, growth and development. An update to the 2007 review in *Annals of Botany*. *Annals of Botany*, *111*(6), 1021–1058. DOI: 10.1093/aob/mct067

Wei, L., Zhang, J., Wang, C., & Liao, W. (2020). Recent progress in the knowledge on the alleviating effect of nitric oxide on heavy metal stress in plants. *Plant Physiology and Biochemistry : PPB*, *147*, 161–171. DOI: 10.1016/j.plaphy.2019.12.021

Weisany, W., Razmi, J., Eshaghabadi, A. H., & Pashang, D. (2024). Silicon nanoparticles (SiNP): A novel and sustainable strategy for mitigating environmental stresses in plants. *Journal of Soil Science and Plant Nutrition*, *24*(2), 2167–2191. DOI: 10.1007/s42729-024-01790-1

Wei, T., Li, X., Yashir, N., Li, H., Sun, Y., Hua, L., Ren, X., & Guo, J. (2021). Effect of exogenous silicon and methyl jasmonate on the alleviation of cadmium-induced phytotoxicity in tomato plants. *Environmental Science and Pollution Research International*, *28*(37), 51854–51864. DOI: 10.1007/s11356-021-14252-3

Weldeslassie, T., Naz, H., Singh, B., & Oves, M. (2017). Chemical contaminants for soil, air and aquatic ecosystem. In *Modern Age Environmental Problems and their Remediation* (pp. 1–22). Springer., DOI: 10.1007/978-3-319-64501-8_1

Witkowska, D., Słowik, J., & Chilicka, K. (2021). Heavy metals and human health: Possible exposure pathways and the competition for protein binding sites. *Molecules (Basel, Switzerland)*, *26*(19), 6060. DOI: 10.3390/molecules26196060

Woldetsadik, D., Drechsel, P., Keraita, B., Itanna, F., & Gebrekidan, H. (2018). Farmers' perceptions on irrigation water contamination, health risks and risk management measures in prominent wastewater-irrigated vegetable farming sites of Addis Ababa, Ethiopia. *Environment Systems & Decisions*, *38*(1), 52–64. DOI: 10.1007/s10669-017-9665-2

Wu, B., Zeng, Z., Wu, X., Li, Y., Wang, F., Yang, J., & Li, X. (2022). Jasmonic acid negatively regulation of root growth in Japonica rice (*Oryza sativa* L.) under cadmium treatment. *Plant Growth Regulation*, *98*(3), 651–667. DOI: 10.1007/s10725-022-00897-8

Wu, F., & Zhang, G. (2002). Alleviation of cadmium-toxicity by application of zinc and ascorbic acid in barley. *Journal of Plant Nutrition*, *25*(12), 2745–2761. DOI: 10.1081/PLN-120015536

Wu, R., Wen, Y., Xu, T., Du, Y., Qi, D., Chang, W., Li, K., Ping, Y., Fan, X., & Song, F. (2025). Rhizophagus irregularis combined with biochar enhances Cr tolerance in Ricinus communis by altering host plant's antioxidant and photosynthetic characteristics. *Industrial Crops and Products*, *227*, 120769. DOI: 10.1016/j.indcrop.2025.120769

Wu, Y., Tan, Z., Zhou, T., Wang, Z., Xue, Z., Yan, Y., He, H., & Pan, J. (2025). Regulation mechanisms of biochar-derived endogenous substances on heavy metal mobility in the soil-crop system: A critical review. *Environmental Research*, *285*, 122557. DOI: 10.1016/j.envres.2025.122557

Wu, Y., Wang, X., Zhang, L., Zheng, Y., Liu, X., & Zhang, Y. (2023). The critical role of biochar to mitigate the adverse impacts of drought and salinity stress in plants. *Frontiers in Plant Science*, *14*(May), 1–22. DOI: 10.3389/fpls.2023.1163451

Xie, Q., Wei, X., Liu, Y., Han, F., & Li, Z. (2022). Germplasm enhancement and identification of loci conferring resistance against Plasmodiophora brassicae in broccoli. *Genes*, *13*(9), 1600. DOI: 10.3390/genes13091600 PMID: 36140766

Xiong, J., Fu, G., Tao, L., & Zhu, C. (2010). Roles of nitric oxide in alleviating heavy metal toxicity in plants. *Archives of Biochemistry and Biophysics*, *497*(1-2), 13–20. DOI: 10.1016/j.abb.2010.02.014

Xu, N., Zhang, N., Yi, P., Chen, L., Dai, H., Zhang, J., Li, W., Li, R., Liu, A., Zhou, Z., & Tu, X. (2024). Integrated physio-biochemistry and RNA-seq revealed the mechanism underlying biochar-mediated alleviation of compound heavy metals (Cd, Pb, As) toxicity in cotton. *Ecotoxicology and Environmental Safety*, *284*, 116974. DOI: 10.1016/j.ecoenv.2024.116974

Yadav, A., Hasan, M., Kour, S., Chandra, R., Singh, V., Noor, I. M., Yahya, M. Z. A., Singh, H. M., & Tripathi, A. (2025). Heavy metal contamination in plants: Sources, monitoring, and data-driven insights. *Macromolecular Symposia*, *414*(1), 2400239. DOI: 10.1002/masy.202400239

Yadav, S. K. (2010). Heavy metals toxicity in plants: An overview on the role of glutathione and phytochelatins in heavy metal stress tolerance of plants. *South African Journal of Botany*, *76*(2), 167–179. DOI: 10.1016/j.sajb.2009.10.007

Yan, G., Huang, Q., Zhao, S., Xu, Y., He, Y., Nikolic, M., Nikolic, N., Liang, Y., & Zhu, Z. (2024). Silicon nanoparticles in sustainable agriculture: Synthesis, absorption, and plant stress alleviation. *Frontiers in Plant Science*, *15*, 1393458. DOI: 10.3389/fpls.2024.1393458 PMID: 38606077

Yang, L., Ji, J., Harris-Shultz, K. R., Wang, H., Wang, H., Abd-Allah, E. F., Luo, Y., & Hu, X. (2016). The dynamic changes of the plasma membrane proteins and the protective roles of nitric oxide in rice subjected to heavy metal cadmium stress. *Frontiers in Plant Science*, *7*, 190. DOI: 10.3389/fpls.2016.00190

Yang, W., Dai, H., Wei, S., & Skuza, L. (2024). The effect of exogenous plant growth regulators on elevated Cd phytoremediation by Solanum nigrum L. in contaminated soil. *Environmental Science and Pollution Research International*, *31*(3), 3964–3975. DOI: 10.1007/s11356-023-31420-9

Yan, Z., Zhang, W., Chen, J., & Li, X. (2015). Methyl jasmonate alleviates cadmium toxicity in *Solanum nigrum* by regulating metal uptake and antioxidative capacity. *Biologia Plantarum*, *59*(2), 373–381. DOI: 10.1007/s10535-015-0491-4

Yap, C. K., Razali, A., Nulit, R., Peng, S. H. T., Yap, C. W., Okamura, H., & Ismail, M. S. (2021). *Biomonitoring of heavy metals in the guava (Psidium guajava) for their health risk assessment in Kluang*. Food Science and Engineering.

Yasin, M. U., Haider, Z., Munir, R., Zulfiqar, U., Rehman, M., Javaid, M. H., Ahmad, I., Nana, C., Saeed, M. S., Ali, B., & Gan, Y. (2024). The synergistic potential of biochar and nanoparticles in phytoremediation and enhancing cadmium tolerance in plants. *Chemosphere*, *354*, 141672. DOI: 10.1016/j.chemosphere.2024.141672 PMID: 38479680

Yeşilyurt, S. (2023). Phytoremediation method and Brassica family: Removal of chromium, cadmium and lead accumulation with broccoli (Brassica oleracea var. italica). *Results in Chemistry*, *6*, 101005. DOI: 10.1016/j.rechem.2023.101005

Yordanova, R., Maslenkova, L., Paunova, S. and Popova, L., 2009. Sensitivity of photosynthetic apparatus of pea plants to heavy metal stress. *Biotechnology & Biotechnological Equipment, 23*(sup1), pp.347-350.

Yousaf, N., Sardar, M. F., Ishfaq, M., Yu, B., Zhong, Y., Zaman, F., Zhang, F., & Zou, C. (2024). Insights in to iron-based nanoparticles (hematite and magnetite) improving the maize growth (Zea mays L.) and iron nutrition with low environmental impacts. *Chemosphere*, *362*, 142781. DOI: 10.1016/j.chemosphere.2024.142781 PMID: 38972262

Yruela, I. (2013). Transition metals in plant photosynthesis. *Metallomics : Integrated Biometal Science*, *5*(9), 1090–1109. DOI: 10.1039/c3mt00086a PMID: 23739807

Yuan, P., Wang, J., Pan, Y., Shen, B., & Wu, C. (2019). Review of biochar for the management of contaminated soil: Preparation, application and prospect. *The Science of the Total Environment*, *659*, 473–490. DOI: 10.1016/j.scitotenv.2018.12.400

Yu, H., Li, W., Liu, X., Song, Q., Li, J., & Xu, J. (2024). Physiological and molecular bases of the nickel toxicity responses in tomato. *Stress Biology*, *4*(1), 25. DOI: 10.1007/s44154-024-00162-0

Yu, X., Yang, L., Fan, C., Hu, J., Zheng, Y., Wang, Z., Liu, Y., Xiao, X., Yang, L., Lei, T., Jiang, M., Jiang, B., Pan, Y., Li, X., Gao, S., & Zhou, Y. (2023). Abscisic acid (ABA) alleviates cadmium toxicity by enhancing the adsorption of cadmium to root cell walls and inducing antioxidant defense system of Cosmos bipinnatus. *Ecotoxicology and Environmental Safety*, *261*, 115101. DOI: 10.1016/j.ecoenv.2023.115101

Yu, X., Zhang, W., Zhang, Y., Zhang, X., Lang, D., & Zhang, X. (2018). The roles of methyl jasmonate to stress in plants. *Functional Plant Biology : FPB*, *46*(3), 197–212. DOI: 10.1071/FP18106

Zaid, A., Ahmad, B., Jaleel, H., Wani, S. H., & Hasanuzzaman, M. (2020). A critical review on iron toxicity and tolerance in plants: role of exogenous phytoprotectants. *Plant micronutrients: Deficiency and toxicity management*, pp.83-99.

Zeb, A., Liu, W., & Zhang, Y. (2022). Effects of nanoparticles on alleviating phytotoxicity of soil heavy metals: potential for enhancing phytoremediation. In *Sustainable Agriculture Reviews 53: Nanoparticles: A New Tool to Enhance Stress Tolerance* (pp. 327–344). Springer International Publishing.

Zeeshan, N., Farooqi, Z. U. R., Ahmad, I., Murtaza, G., Jamal, A., Saifullah, , Abdul Qadir, A., & Radicetti, E. (2023). Trace Metals in Rice Grains and Their Associated Health Risks from Conventional and Non-Conventional Rice Growing Areas in Punjab-Pakistan. *Sustainability (Basel), 15*(9), 7259. DOI: 10.3390/su15097259

Zhang, C., Huang, R., Zhan, N., & Qin, L. (2023). Methyl jasmonate and selenium synergistically mitigative cadmium toxicity in hot pepper (*Capsicum annuum* L.) plants by improving antioxidase activities and reducing Cd accumulation. *Environmental Science and Pollution Research International, 30*(34), 82458–82469. DOI: 10.1007/s11356-023-28273-7

Zhang, D., Wang, H., Zhang, Y., Su, Z., Hu, T., Liu, J., Ding, Q., Niu, N., & Ma, L. (2024). Methyl jasmonate enhances the safe production ability of Cd-stressed wheat by regulating the antioxidant capacity, Cd absorption, and distribution in wheat. *Plant Physiology and Biochemistry : PPB, 212*, 108788. DOI: 10.1016/j.plaphy.2024.108788

Zhang, J., Zeng, L., Sun, H., Wu, H., & Chen, S. (2017). Adversity stress-related responses at physiological attributes, transcriptional and enzymatic levels after exposure to Cu in Lycopersicum esculentm seedlings. *Scientia Horticulturae, 222*, 213–220. DOI: 10.1016/j.scienta.2017.05.027

Zhang, K., Wang, G., Bao, M., Wang, L., & Xie, X. (2019). Exogenous application of ascorbic acid mitigates cadmium toxicity and uptake in maize (*Zea mays* L.). *Environmental Science and Pollution Research International, 26*(19), 19261–19271. DOI: 10.1007/s11356-019-05265-0

Zhang, L., Liu, Z., Song, Y., Sui, J., & Hua, X. (2024). Advances in the involvement of metals and metalloids in plant defense response to external stress. *Plants*, *13*(2), 313. DOI: 10.3390/plants13020313

Zhang, X., Zhao, Y., Huang, L., Luo, X., Zhang, C., Mao, Z., Yang, H., & Wang, X. (2025). Zinc oxide nanoparticles alleviated Cd toxicity in Hibiscus syriacus L. by reducing Cd translocation and improving plant growth and root cellular ultrastructure. *Journal of Hazardous Materials*, *491*, 137920. DOI: 10.1016/j.jhazmat.2025.137920 PMID: 40086238

Zhang, Y., Wang, R., Wang, X., Zhao, C., Shen, H., & Yang, L. (2023). Nitric oxide regulates seed germination by integrating multiple signalling pathways. *International Journal of Molecular Sciences*, *24*(10), 9052. DOI: 10.3390/ijms24109052

Zhang, Y., Yang, C., Liu, S., Xie, Z., Chang, H., & Wu, T. (2024). Phytohormones-mediated strategies for mitigation of heavy metals toxicity in plants focused on sustainable production. *Plant Cell Reports*, *43*(4), 99. DOI: 10.1007/s00299-024-03189-9

Zhao, D., Wang, P., & Zhao, F. J. (2024). Toxic metals and metalloids in food: Current status, health risks, and mitigation strategies. *Current Environmental Health Reports*, *11*(4), 468–483. DOI: 10.1007/s40572-024-00462-7 PMID: 39352604

Zhao, L., Lu, L., Wang, A., Zhang, H., Huang, M., Wu, H., Xing, B., Wang, Z., & Ji, R. (2020). Nano-biotechnology in agriculture: Use of nanomaterials to promote plant growth and stress tolerance. *Journal of Agricultural and Food Chemistry*, *68*(7), 1935–1947. DOI: 10.1021/acs.jafc.9b06615 PMID: 32003987

Zhao, S., Zhao, S., & Wang, B. (2025). Combined application of biochar and PGPB on crop growth and heavy metals accumulation: A meta-analysis. *Environmental Pollution*, *381*, 126626. DOI: 10.1016/j.envpol.2025.126626

Zhao, Y., Wang, J., Huang, W., Zhang, D., Wu, J., Li, B., Li, M., Liu, L., & Yan, M. (2023). Abscisic-acid-regulated responses to alleviate cadmium toxicity in plants. *Plants*, *12*(5), 1023. DOI: 10.3390/plants12051023

Zhou, M. L., Yang, X. B., Zhang, Q., Zhou, M., Zhao, E. Z., Tang, Y. X., Zhu, X. M., Shao, J. R., & Wu, Y. M. (2013). Induction of annexin by heavy metals and jasmonic acid in *Zea mays. Functional & Integrative Genomics, 13*(2), 241–251. DOI: 10.1007/s10142-013-0316-5

Zhu, J. K. (2002). Salt and drought stress signal transduction in plants. *Annual Review of Plant Physiology and Plant Molecular Biology, 53*(1), 247–273. DOI: 10.1146/annurev.arplant.53.091401.143329 PMID: 12221975

Zhu, Y., Xu, F., Liu, Q., Chen, M., Liu, X., Wang, Y., Sun, Y., & Zhang, L. (2019). Nanomaterials and plants: Positive effects, toxicity and the remediation of metal and metalloid pollution in soil. *The Science of the Total Environment, 662*, 414–421. DOI: 10.1016/j.scitotenv.2019.01.234 PMID: 30690375

Zong, H., Liu, J., Wang, F., & Song, N. (2020). Root morphological response of six peanut cultivars to chromium (VI) toxicity. *Environmental Science and Pollution Research International, 27*(15), 18403–18411. DOI: 10.1007/s11356-020-08188-3

Zulfiqar, B., Raza, M. A. S., Saleem, M. F., Aslam, M. U., Iqbal, R., Muhammad, F., Amin, J., Ibrahim, M. A., & Khan, I. H. (2022). Biochar enhances wheat crop productivity by mitigating the effects of drought: Insights into physiological and antioxidant defense mechanisms. *PLoS ONE, 17*(4 April), 1–16. DOI: 10.1371/journal.pone.0267819

Zulfiqar, A., Gul, B., Saleem, A., Islam, A., Zulfiqar, U., Ali, M. F., Nawaz, M., Al-Ghamdi, A. A., & Rizwana, H. (2024). Salicylic Acid Mitigates Cadmium Stress in Wheat: Experimental Insights Into Growth and Biochemical Parameters. *Scientifica, 2024*(1), 6887694. DOI: 10.1155/sci5/6887694

Zulfiqar, F., & Ashraf, M. (2021). Bioregulators: Unlocking their potential role in regulation of the plant oxidative defense system. *Plant Molecular Biology, 105*(1), 11–41. DOI: 10.1007/s11103-020-01077-w

Zulfiqar, U., Jiang, W., Xiukang, W., Hussain, S., Ahmad, M., Maqsood, M. F., Ali, N., Ishfaq, M., Kaleem, M., Haider, F. U., Farooq, N., Naveed, M., Kucerik, J., Brtnicky, M., & Mustafa, A. (2022). Cadmium phytotoxicity, tolerance, and advanced remediation approaches in agricultural soils; a comprehensive review. *Frontiers in Plant Science*, *13*, 773815. DOI: 10.3389/fpls.2022.773815

# About the Editor

**Riaz Ahmad** completed his Ph.D. at Department of Horticulture, Bahauddin Zakariya University, Multan, Pakistan. Currently, serving as Lecturer at The University of Agriculture, Dera Ismail Khan, Pakistan. He is working on the characterization of horticultural crops through morphological, biochemical, and molecular approaches. Moreover, he is also working in the area of abiotic stress tolerance through phytohormones, biostimulants, and omics strategies in horticulture crops. He is an expert researcher in the area of abiotic stress such as salinity, drought, and heavy metals. Apart from this, he also has expertise in the use of plant leaf extract on numerous horticulture crops against abiotic stress tolerance focusing on higher yield. His findings have generated more than 60 publications in national and international peer-reviewed journals.

# Index

www.ingramcontent.com/pod-product-compliance
Lightning Source LLC
LaVergne TN
LVHW060055220926
848484LV00006BA/1470